CULTURES OF COMMODITY BRANDING

Publications of the
Institute of Archaeology, University College London

Series Editor: Ruth Whitehouse
Director of the Institute: Stephen Shennan
Founding Series Editor: Peter J. Ucko

The Institute of Archaeology of University College London is one of the oldest, largest, and most prestigious archaeology research facilities in the world. Its extensive publications programme includes the best theory, research, pedagogy, and reference materials in archaeology and cognate disciplines through publishing exemplary work of scholars worldwide. Via its publications, the Institute brings together key areas of theoretical and substantive knowledge, improves archaeological practice, and brings archaeological findings to the general public, researchers, and practitioners. It also publishes staff research projects, site and survey reports, and conference proceedings. The publications programme, formerly developed in-house or in conjunction with UCL Press, is now produced in partnership with Left Coast Press, Inc. The Institute can be accessed online at http://www.ucl.ac.uk/archaeology.

Recent Titles
Andrew Bevan and David Wengrow (eds.), ***Cultures of Commodity Branding***
Peter Jordan (ed.), ***Landscape and Culture in Northern Eurasia***
Peter Jordan and Marek Zvelebil (eds.), ***Ceramics before Farming***
Marcos Martinón-Torres and Thilo Rehren (eds.), ***Archaeology, History, and Science***
Miriam Davis, ***Dame Kathleen Kenyon***
Elizabeth Pye (ed.), ***The Power of Touch***
Russell McDougall and Iain Davidson (eds.), ***The Roth Family, Anthropology, and Colonial Administration***
Eleni Asouti and Dorian Q. Fuller, ***Trees and Woodlands of South India***
Tony Waldron, ***Paleoepidemiology***
Janet Picton, Stephen Quirke, and Paul C. Roberts (eds.), ***Living Images***
Timothy Clack and Marcus Brittain (eds.), ***Archaeology and the Media***
Sue Colledge and James Conolly (eds.), ***The Origins and Spread of Domestic Plants in Southwest Asia and Europe***
Gustavo Politis, ***Nukak***
Sue Hamilton, Ruth Whitehouse, and Katherine I. Wright (eds.), ***Archaeology and Women***
Andrew Gardner, ***An Archaeology of Identity***
Barbara Bender, Sue Hamilton, and Chris Tilley, ***Stone World***
James Graham-Campbell and Gareth Williams (eds.), ***Silver Economy in the Viking Age***
Gabriele Puschnigg, ***Ceramics of the Merv Oasis***
Joost Fontein, ***The Silence of Great Zimbabwe***

Critical Cultural Heritage Series, Beverley Butler (ed.)
Katharina Schramm, ***African Homecoming***
Mingming Wang, ***Empire and Local Worlds***
Dean Sully (ed.), ***Decolonizing Conservation***
Ferdinand de Jong and Michael Rowlands (eds.), ***Reclaiming Heritage***
Beverley Butler, ***Return to Alexandria***

CULTURES OF COMMODITY BRANDING

Andrew Bevan
David Wengrow

Editors

Routledge
Taylor & Francis Group
LONDON AND NEW YORK

First published 2010 by Left Coast Press, Inc.

Published 2016 by Routledge
2 Park Square, Milton Park, Abingdon, Oxon OX14 4RN
711 Third Avenue, New York, NY 10017, USA

Routledge is an imprint of the Taylor & Francis Group, an informa business

Library of Congress Cataloging-in-Publication Data

Cultures of commodity branding / Andrew Bevan, David Wengrow, editors.
p. cm.—(Publications of the Institue of Archaeology, University College London)
Includes index.
ISBN 978-1-59874-541-2 (hardcover : alk. paper)
1. Material culture—History. 2. Manufactures—History. 3. Marks of origin—History. 4. Trademarks—History. 5. Branding (Marketing)—History.
I. Bevan, Andrew, 1974- II. Wengrow, D.
GN430.C856 2010
306.4′609—dc22

2010013286

ISBN 978-1-59874-541-2 hardcover

Contents

List of Illustrations

Acknowledgements

The present volume arises from an international conference held at University College London in 2009, which assembled researchers in archaeology, anthropology, and marketing to consider the role of branding in a broader chronological and cross-cultural perspective than is usually attempted. The conference was funded by grants from the British Academy and the Institute of Archaeology (UCL), whose support is gratefully acknowledged. We would also like to thank all of the original participants for their contributions. In addition to those whose work appears in this volume, they include: John Bennet, Alex Bentley, Fabian Csaba, Güliz Ger, Douglas Holt, Olga Kravets and Özlem Sandikçi. Danny Miller, Anne Pyburn, Mike Rowlands, Stephen Shennan, Sue Sherratt, and Rick Wilk are also to be thanked for their support in putting the conference together or for acting as panel chairs and discussants. We are grateful to Michelle De Gruchy, Sarah Foster, and Mary Shepperson for their assistance in running the conference and to Graeme Were and Haidy Geismar (Material World Blog) and Ruth Mortimer (Brand Strategy) for help with publicity. A final thanks to the copy editor and anonymous reviewers of this volume for their extremely useful suggestions.

Chapter 1

Introduction

Commodity Branding in Archaeological and Anthropological Perspectives

David Wengrow

The origins of this book lie in coincidence. About 5 years ago I was reading Naomi Klein's *No Logo* (2000), with no special academic interest in the world of branding. At much the same time, I was teaching graduate classes at UCL's Institute of Archaeology, some of which dealt with the early development of writing, pictorial art, and urban economies in the Middle East. UCL itself was about to undergo an expensive and controversial rebranding exercise. A new university logo was presented to staff, and instructions were provided on how to deploy it, what it represented, and how this was to be reflected in our professional lives and public conduct. My response to this set of circumstances eventually came together in an article ("Prehistories of Commodity Branding") published in 2008 in *Current Anthropology*, together with comments—some very favourable, others highly critical—from an unusual combination of scholars in archaeology, anthropology, art history, and marketing. Its main argument, which I summarise below, was (in brief) that something we would recognise today as commodity branding can be found in much older societies, including those of ancient Egypt and Mesopotamia, where the first cities and large-scale economies emerged around 6,000 years ago.

Within months of its publication, the findings of the article had been reported in a number of major media outlets, such as *The Telegraph* and *New Scientist*. I gave a live radio interview about the origins of branding and for a short time was contacted on an almost daily basis by journalists wanting to write about branding in the ancient world. Summaries of the article, many of which exaggerated and/or oversimplified its claims, began to appear in consumer blogs based in such diverse places as India and Poland. A leading business think-tank invited me to a closed debate on the future of global branding, hosted by a well-known TV pundit, where I found myself a lone archaeologist in the company of CEOs from Interbrand and Coca-Cola. Although this was all quite flattering, and fully in line with the new brand image of my university, it was also confusing. I had not, contrary to what some people wrote or said, made any great new "discovery" (a number of my interviewers were disappointed when I explained that I could not provide them with an image of "the ancient brand" I had found). Rather, I had simply interpreted information that is quite familiar in my field in a new way. So, why the fuss?

The reason seemed quite clear, but also problematic. Most people are aware of and talk about commodity branding on the assumption that it is the product of modern capitalist markets, and that we therefore know quite intuitively what it involves. Of course, there are marketing experts and brand gurus, some of whom I have now met, who know far more about how it works than others, but it has generally been assumed that we, as consumers of brands, and they, as designers of brands, are all talking about roughly the same thing. It is also clear that capitalist logo has become an emotive subject, one that draws fire from a variety of moral-ethical standpoints, and is not infrequently a target of grass-roots activism. Much of this critique is directed at "brand bullies"—global multinationals—rather than at the phenomenon of branding itself, but the line often seems blurred. There is a widespread perception that the branding of things, people, and knowledge is a distinctive creation of the postindustrial West, which is now being exported around the world, leading to the erosion of cultural diversity and local identities in new and unprecedented ways.

Viewed against this backdrop, my article was clearly making some unusual claims. Specifically, I was trying to separate the history of branding from the history of capitalism and to locate its origins outside the recent history of Western societies. As an anonymous reviewer of my *Current Anthropology* submission pointed out (and as I had initially failed to realise), this had already been done some years earlier, by social historians Gary Hamilton and Chi-kong Lai. In their contribution to *The Social Economy of Consumption* (1989), Hamilton and Lai convincingly argued that a complex system of commodity branding—applied to goods such as rice, tea, wine, scissors, and medicines—had existed in late imperial China, where it can be traced back to the Song dynasty of the 10th century AD (see, more recently Hamilton 2006; also Eckhardt and Bengtsson 2009, but note their incorrect dating of the non-Chinese material). Their basic claim—that branding was neither uniquely "modern," "Western," nor "capitalist"—resonated with my own negative response to Naomi Klein's description of the "origins" of branding in postindustrial Europe. It seemed to me that, in its fundamentals, what Klein was describing could be quite easily traced across much earlier periods of human history: the anonymity of mass consumption, the shift of corporate enterprise from creating novel products to image-based systems of distinction, and the emergence of generic labels that attach (often highly imaginative) biographies to otherwise indistinguishable goods and packaging.

The Brand: Between Marxist Critique and Market-Speak

All of this was quite familiar from my earlier research into the origins of urban societies and writing systems in the Middle East, between around 4000 and 3000 BC. For instance, many examples of what is regarded as the earliest Egyptian writing appear on labels attached to a wide range of commodities such as textiles, oils, and alcoholic drinks that were stored and transported in standardised ceramic containers. These labels not only provide bureaucratic information relating to quality and provenance, but also carry images that link the commodities in question to royal ceremonies and to particular locations in which the king performed rituals to ensure the fertility of the land (Wengrow 2006: 198–207, 2008: 9–10).

A broadly comparable system of product marking can be documented in ancient Mesopotamia (today's Iraq), where a series of administrative terms (appearing in texts of the 3rd millennium BC) are used to convey information concerning: "(1) emblems or symbols of deities, households, or individuals, (2) marking tools bearing these emblems or symbols, (3) marks which were the result of applying these tools to animals, human beings or boats and, finally, (4) verbs describing these marking activities" (de Maaijer 2001: 301). Daniel Foxvog (1995) actually translates one such term—which denotes the marking of cattle and other commodities—as "to brand."

Though clearly of little value when considered in isolation, such instances point towards the possibility that temple and palace bureaucracies in the ancient Near East may have been using systems of marking and notation, not only to monitor the quantity and quality of manufactured goods (as has long been recognised) but also to enhance the value of such products through specialised procedures of packaging and labelling. With such considerations in mind, I began to read more deeply into the history of critical theory and semiotics: mid-20th-century theoretical movements that have become highly unfashionable in the social sciences but which at least offered a starting point for thinking in general terms about branding and advertising as cultural phenomena. What I found there were still stronger claims, formulated within a broadly Marxian perspective, for the exclusively modern character of branding and its transformative effects in postindustrial societies.

Both the German (Frankfurt) and French (semiotic) schools seemed to implicate branded commodities in the decline of modes of subjectivity based on kinship, class relations, and caste, arguing that mass consumption creates a new set of normative identities, tying consumers to the exploitative conditions of capitalist production (e.g., Barthes 1977 [1964]; Horkheimer and Adorno 1996 [1944]). Jean Baudrillard (1968, 1970, 1981), for example, saw modern branding practices as a form of cultural alchemy specific to capitalist modernity and unparalleled in earlier social formations. The brand sign, he argued, brings together in an ephemeral material form two conflicting psychological tendencies: the drive for short-term gratification and the long-term need for transcendence—what we might call, for brevity's sake, the "*Coca-Cola Is Life* effect." The intended outcome is a short-lived transcendence of self that can only be sustained through further acts of purchase and consumption, so that commodity branding ultimately seeks to create a whole pattern of social and economic dependency. More recent critiques, including those offered by anthropologists, tend to reinforce the association of branding with "late capitalism" (e.g., Carrier 1995; Wernick 1991). Even analyses of non-Western advertising (e.g., in contemporary Japan; see Tobin 1992) take Western marketing strategies as their touchstone of comparison, rather than pursuing possible continuities with local or indigenous forms of branding.

Turning to the cultural wing of modern marketing literature, I was surprised to find many overlaps with the critical, Marxian tradition of brand analysis. Although the ancient roots of product labelling are often whimsically mentioned in this literature (with a nod to the ancient Greeks and such things as potters' marks), branding itself is clearly regarded by many marketing professionals (and professors of marketing) as having only recently become a major player on the field of social change (e.g., Chevalier and Mazzalovo 2004; Holt 2006, 2008; for an exception, see Moore and Reid 2008 and comments below). I was particularly struck by how some marketing analysts had picked up on the earlier psychoanalytical literature, arguing that the appeal of brands is rooted in the "quick fixes" they offer for deep existential crises in modern societies, generating distinct and self-perpetuating forms of dependency between consumers and their most coveted products (e.g., Holt 2002: 87, 2004). The major difference seemed to be that 21st-century experts on marketing treat the construction of self-image through mass consumption as a basis for commercial strategies that, if properly managed and targeted, might transform an ailing business or public institution into a global success story. Their 20th-century ancestors, by contrast, thought the same material-psychological drives were destroying the moral fabric of society in general. Even Raymond Williams's (1980 [1960]) characterisation of marketing as a modern form of "magic," "symbolism," and "mythology," I discovered, has been enthusiastically incorporated into professional brand-speak, but as an endorsement of good practice rather than a form of psychological warfare waged on the weak by the powerful (e.g., Roberts 2004).

In short, it was becoming clear to me that arguments for the modernity of commodity branding have been widely mounted (both within and outside academia), often from very diverse standpoints, and in the service of very different social agendas. But what happens to these various, entrenched positions on contemporary brand culture if the West actually turns out to be a relative latecomer in its adoption of systematic product branding? What are the implications of decoupling brand economies, as social and historical phenomena, from capitalist modes of production? What happens if we elect to see contemporary Western capitalism as just one context (or better, set of contexts) in which branding—a phenomenon common to large-scale societies since before the time of the Great Pyramids—has taken on new forms and cultural permutations? Before exploring these questions and some of the answers offered in this volume, it seems important to summarise the evidence presented in "Prehistories of Commodity Branding" (Wengrow 2008).

Prehistories of Commodity Branding

In a seminal piece on the social construction of value, the anthropologist Igor Kopytoff (1986) argued that commodities are not things in themselves but rather situational entities: elements of material culture (or persons) that have become temporarily enmeshed within a particular set of social and moral codes; codes that determine the character and pace of their circulation. Commodity situations are those in which the exchangeability of one thing for another becomes, for a time, its most socially salient feature. In the same work, Kopytoff also observed that the production of commodities is "a cultural and cognitive process: commodities must not only be produced materially as things, but also culturally marked as being a certain kind of thing" (p. 64). It is this latter observation that has remained underdeveloped in anthropological studies of value and exchange, a neglect that finds its counterpart in recent archaeological theories of commoditisation (e.g., Renfrew 2001, 2005) that pay little attention to concrete mechanisms of sealing, labelling, and marking.

Specialised forms of product labelling in premodern societies have tended to be studied in one of two ways. They may be considered in terms of their administrative roles (i.e., their use in monitoring the circulation of people and goods—an approach usually taken by archaeologists and anthropologists). Or they are studied from a purely iconographic perspective, with various kinds of labels treated as art objects, divorced from the commodities to which they were attached, and hence from the world of social transactions (an approach usually followed by art historians; but for a notable exception, see Winter 2000). My primary aim in "Prehistories of Commodity Branding" was to question this artificial separation of functions (and of modern disciplinary interests) as well as the wider reconstructions of ancient economy and society that it supports. I attempted to do this through a detailed study of how commodities were sealed and marked in Mesopotamia during the transition from village to urban life (*ca.* 7000–3000 BC).

The kind of sealing practices with which I was concerned involve two basic physical components: (1) the seal itself, often a highly crafted object made of some hard substance, which carries an intaglio design that is impressed by rolling or stamping; and (2) the sealing, made of a soft adhesive substance such as clay or wax, which serves the dual functions of receiving the seal image and holding shut the closure of a container through its attachment over the lip (Figure 1.1, centre). Sealing practices of this kind are no longer a standard feature of commercial life. But their replacement by industrial adhesives and synthetic wrapping materials is a relatively recent development, and mass-produced ("skeuomorphic") replicas of seal impressions continue to be featured on labels for a wide range of modern commodities, notably those—such as alcoholic drinks, perfumes, dairy products, and other comestibles—that must be sealed for practical as well as social reasons, to avoid spoiling through exposure. This modern (symbolic) use of seal imagery to evoke nostalgia and authenticity is perhaps responsible, in part, for disguising the fact that sealing practices of this kind, when viewed from a historical perspective, constitute the

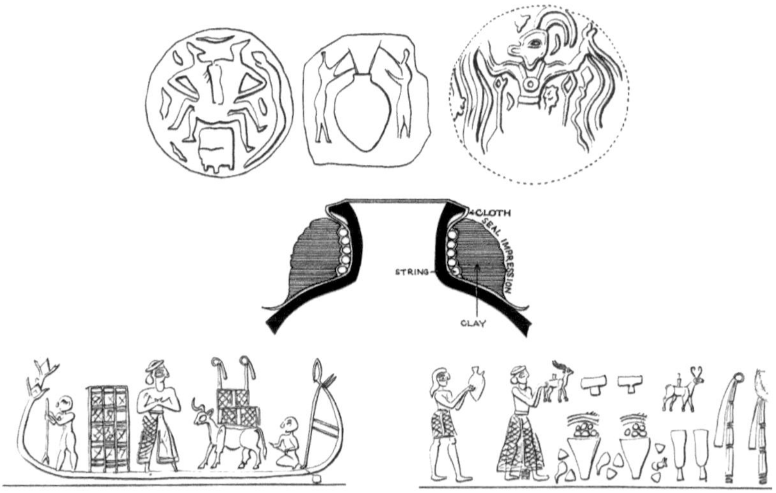

Figure 1.1. Images impressed onto the clay sealings of commodity containers in early Mesopotamia and south-west Iran. (above: stamped impressions, 5th–early 4th millennium BC; below: rolled impressions, 4th millennium BC), and an illustration of the sealing mechanism (centre).

earliest-known technique for mechanically reproducing a crafted image (and note their striking absence from Walter Benjamin's famous [1999 (1936)] essay on "the work of art in the age of mechanical reproduction").

In the alluvial plains between the Tigris and Euphrates Rivers in Mesopotamia, the practice of sealing containers with impressed images in this manner can be traced back to the late 7th millennium BC, some 2,000 years prior to the emergence of cities in the same region. The function of sealing practices in these early village economies has been much debated (see, e.g., Ferioli ed. 1994). What role did they play in prehistoric communities that might account for their rapid spread across the plains of northern Mesopotamia? As we are dealing here with "household economies," it seemed useful to approach the question via anthropological studies of wrapping and sealing practices in more recent domestic settings.

James Carrier (1993), for instance, has considered how wrapping gifts, tying bows, and attaching greetings works to overcome a contradiction between the generic qualities of (modern) branded products and the social requirements of ceremonial exchange in contemporary American households. Here, domestic "sealing practices"—despite their reliance on industrially produced wrapping paper, adhesives, and greetings cards—still work to subvert the commodity status of mass-produced goods, recalling Pierre Bourdieu's (1977: 71) slightly cynical observation that gift exchanges can be viewed as commodity relations, slowed to a snail's pace and filtered through a series of reciprocal performances that deny the self-interested nature of the transaction. To invert Kopytoff's formulation, in a world saturated by branded commodities, it is *gifts* that must be culturally marked as being "a certain kind of thing."

The Neolithic villages in which specialised sealing practices first developed present an interesting contrast. The household was a primary locus of production as well as exchange, and material goods were often highly individualised objects with personal histories of manufacture and ownership. This applies to the stone seals themselves, which might be equally well understood as personal amulets (Charvát 1994). Each had a unique shape and surface design and was perforated for attachment to the body, suggesting close relationships between particular persons, the seals they carried, and the images distributed from those seals (Figure 1.2). The act of placing a band of wet clay over the mouth of a container and impressing it with a carved stone amulet left a distinguishing mark that could be used to trace the product back to a particular individual or institution: a point of origin. This seemingly innocuous development would have far-reaching consequences, still detectable in today's consumer cultures.

The presence of a clay sealing demonstrated the integrity of the package and its contents, which was particularly important in the case of organic comestibles, and had the potential to reduce the risks involved

Figure 1.2. Seal amulets of the late Neolithic period, from northern Iraq (6th millennium BC).

in exchanges between unfamiliar partners. But it also introduced new possibilities for mystification in the circulation and consumption of commodities. Then, as now, breaking a seal always disturbs a prior set of relationships: between the owner of the sealed object, the owner of the seal used to fasten it, and the agencies evoked by the image carved on the seal's surface, which were sometimes of a supernatural kind. It is therefore both something of a violation and something of a temptation, setting in motion a chain of consequences, the outcomes of which cannot always be foreseen, and may lead to misfortune. Seals have the potential to rewrite social history, and as such have often been viewed as portentous and dangerous objects in their own right. Later Mesopotamian "dream omens" credited them with magical powers, including the power to produce or destroy offspring (Pittman 1995).

Over succeeding millennia, Mesopotamian societies underwent a transition from village to urban life that was unprecedented in human history. An important aspect of that process, described in more detail in my original article, was the standardisation of widespread forms of material culture, of which the most archaeologically visible are ceramics (Figure 1.3). By around 4000 BC, the highly individualised, hand-made and -decorated containers so characteristic of Neolithic village life in this region had disappeared. What replaced them were drab and uniform vessel types, manufactured with the aid of mechanical devices such as the wheel or constructed out of prefabricated, modular components. Seals and sealing practices, too, underwent significant changes across Mesopotamia (Pittman 2001). The general trend was towards standardisation of the seal form as well as a widespread shift from stamping to rolling (and hence from flat to cylindrical seals) as the method of impressing images onto the clay closures of commodity containers.

The invention of the earliest-known writing system (*ca.* 3300 BC), and the chance preservation of early administrative texts at the site of Uruk in southern Iraq, provides detailed insights into the range of commodities and specialised packaging that was available in urban settings by the late 4th millennium (Nissen et al. 1993). Dairy products, meats, and

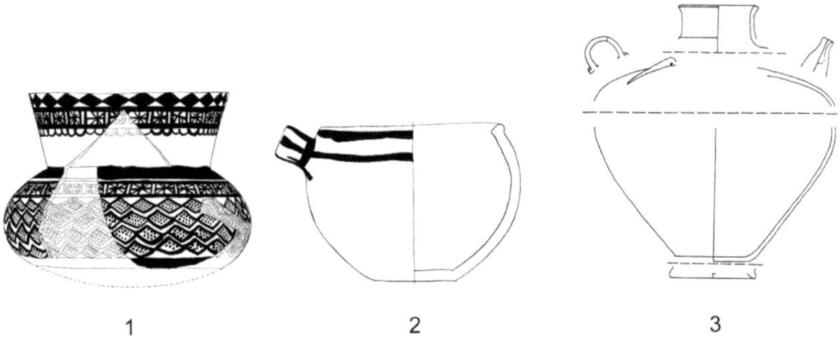

Figure 1.3. Towards mass production: characteristic Mesopotamian vessel types of the 6th (1), 5th (2), and 4th (3) millennia BC, showing a progression from highly ornamented coil- or slab-built to plain wheel-made and modular forms.

oils for cooking and cosmetic use as well as types of bread and beer are enumerated and classified according to fine-grained criteria of distinction, leaving no doubt as to the sophistication of the urban economy and its ability to process and package organic produce on a very large scale. These texts offer a snapshot of innovations in diet and consumption that took many centuries to evolve. As the range of commodities had expanded, notably from the 5th millennium BC onward, the miniature designs impressed onto their clay sealings had also become more vivid and memorable. Amongst them we find a new bestiary of real and fictitious animals in poses of violence or copulation, a panoply of scenes showing people drinking and feasting, and a hybrid figure with human body and ram's head who holds aloft a pair of snakes. This rapidly expanding cast of characters—and the real human agents behind their production—now stood guard over the world of commodities, occupying an important space of the social imagination between novel products and their consumption (Figure 1.1, upper row).

By the time of the earliest writing, standardised commodity containers were sealed with images that conveyed clear narrative themes (often with a strong, masculine component) such as the conquest of enemies, the nurturing of wild animals, and pious presentations of gifts to the gods (Figure 1.1, lower row). Scenes of industry (e.g., potting and weaving) also appear, featuring squatting workers (usually assumed to be female) with distinctive pig-tailed hairstyles. A striking feature of this period is the dissemination of standardised goods over a truly enormous area, reaching across the Mesopotamian alluvium from the head of the Persian Gulf up the headwaters of the Euphrates (in modern-day Turkey) and penetrating the western plains and mountains of Iran. Archaeologists term this the "Uruk

Expansion," after the aforementioned site in southern Mesopotamia, which had reached a size of around 250 ha (with an estimated population of 20,000) by the late 4th millennium BC (Algaze 1993).

The causes of the Uruk Expansion have been widely debated, and scholarly disagreements often hinge on the nature of commodity flows between urban centres on the Mesopotamian alluvium and a surrounding highland periphery (along the Taurus and Zagros ranges). The latter regions were less densely populated but controlled access to metal and mineral resources that were lacking in lowland Mesopotamia (see chapters in Rothman 2001). Andrew Sherratt (2004), in an important contribution, noted that the outward flow of urban products into this wider periphery was accompanied by internal specialisation in primary production and "lengthened chains of transformation" in the manufacture of textiles, alcoholic beverages, and a wide range of foodstuffs. He further suggested that these new consumables were part and parcel of an urban *habitus*—a system of customary social distinctions—transmitted far beyond the heartlands of early urbanisation into less densely populated regions and settlements.

My 2008 article added a further dimension to this reconstruction, by arguing that specialised techniques of commodity marking played a more complex role in the dissemination of this urban lifestyle than had previously been realised. Urban centres, I suggested, not only possessed an unrivalled capacity to monitor the production, and thus guarantee the quality, of mass-produced goods (as testified in early written records); they also invested in standardised systems of product packaging and labelling that altered the nature of commodity flows in their favour (and see further discussion of "brand economies," below). A key feature of that process, as A. Sherratt (1999) had pointed out, was the growing importance of commodities (notably woollen textiles, dairy products, alcoholic beverages, oils, and metals in both finished and ingot forms) that could be produced along a wide spectrum of quality and purity, using complex (and often rare) ingredients, and absorbing the added value of skilled and intensive labour. Such goods were highly susceptible to adulteration and fakery, placing a premium on specialised modes of marking and packaging that ensured quality and signalled relationships of exclusivity across shared scales of sumptuary value (and see Bevan, this volume).

To summarise, sealing practices clearly played multiples role in the earliest urban economies, not all of which can or should be subsumed under the rubric of commodity branding (see Winter 2008). Among those roles are the use of seals as mechanical devices for marking the quality (and perhaps also provenance) of goods, for guaranteeing the pristine condition of perishable foodstuffs, and for tracking the movement of products within bureaucratic systems of resource management. Additionally, sealing practices involved the application to widely traded commodities of striking visual images, images that linked these otherwise anonymous objects to central cultural concerns, including ceremonial transactions that reproduced an overarching order of relations between humans and the sacred. In such contexts, they formed part of a process of (limited) decommoditisation, carried out on what was, by any standards, an industrial scale. It is this process, I was suggesting, that can be best described as the world's first system of commodity branding.

Cultures of Commodity Branding

Picking up where "Prehistories of Commodity Branding" left off, this book is the first concerted attempt to approach commodity branding on a comparative scale, from the disciplinary perspectives of archaeology and anthropology. Unlike previous studies, we approach our topic without any assumed limitations of time, culture, or space, addressing both general and specific features of brand economies and considering the interface between distinct cultures of commodity branding as they come into contact with one another.

Clearly there are dangers, in any such enterprise, of reducing non-Western or premodern forms of branding to shadowy precursors of something apparently familiar. We would contrast our approach, for instance, to that of Moore and Reid (2008) who, writing for a leading business and marketing journal, have recently sought to identify "proto-brands" among ancient sealing devices from the Indus Valley. From this starting point, they go on to hypothesise a long-term evolution of branding practices from marks concerned with origin, authenticity, and the logistical management of goods to the charismatic brands of modern commerce. We posit no such evolutionary scheme here and, as will be apparent by now, view charismatic product marking as an integral feature of premodern as well as modern branding.

The chapters in this volume work cumulatively to question the cultural preconceptions and historical assumptions in which the familiarity of modern brands is grounded. Each treats a particular body of archaeological or ethnographic data in depth and, in spite of their widely varying scales of analysis, all share a number of features that distinguish them as a group from earlier studies of commodity branding. First, they approach branding from the "ground up," in terms of the customary practices and realities of everyday interaction. Their data are not the carefully codified or statistically manageable aspects of commercial relations, through which marketing and business studies have conventionally approached the analysis of brands. They are concerned, instead, with the manner in which brands and branded goods are understood as mediators of social relationships; relationships grounded in wider notions of trust and authenticity, and distinction that vary between cultural and historical contexts. It is no accident, then, that major subthemes to emerge from this collection are those of the "fake brand," the "second-hand brand," and the "misunderstood" or "reinvented" brand, socially reconstituted in ways that are likely to escape the professional concerns of brand managers, designers, and analysts.

The "Brand Economy" and the "Bazaar Economy": Some Definitions

We offer no ready-made formula for the comparative study of branding, so it seems important to clarify at the outset what can often seem a rather vague relationship between "branding" and "commodities." Here, the existing anthropological literature offers some points of departure, although none of them are entirely satisfactory (see Wengrow [2008] for a more detailed review). Perhaps the most directly instructive is Frank Fanselow's (1990) study of a commercial town in Tamil Nadu, southern India, where the coexistence of different outlets specialising in branded and nonbranded commodities provides insights into their different transactional properties. Based on ethnographic observations of how transactions are conducted with these different kinds of goods, Fanselow drew a broad analytical distinction between "brand economies" and "bazaar economies," which is critically used in a number of chapters that follow.

In the "bazaar economy," goods are unbranded and ungraded, so that consumers have little opportunity to assess quality and quantity before purchase. Transactions in goods of this kind often involve the mobilisation of personal networks of loyalty and affiliation between traders and consumers, so that any breach of trust (e.g., by selling faulty or adulterated goods) threatens the integrity of a larger social whole. The logic of the bazaar economy is not particular to any cultural group or region but arises from the ambiguous and heterogeneous nature of the commodities transacted. Hence Fanselow would include European markets in second-hand merchandise under the category of "bazaar economies." Such systems, he argues, are subject to constraints of scale that do not apply to brand economies.

In the "brand economy," goods are standardised, strictly graded, and directly substitutable for one another. Consumers can exchange knowledge about the merits and deficiencies of particular genres of product: which are to be trusted and which avoided. Information circuits and social networks can therefore be generated around particular classes of commodity. The fact that these commodities are (at least nominally) pretested for quality and sold in sealed packaging means that personal loyalty and trust are no longer so important at the point of transaction. Branded goods (and, to some extent, the personnel who deal in them) can then be treated as *types* of thing or *types* of person, and their management can be rationalised accordingly. By virtue of this, brand economies are more able to quantify and transparently display their assets than bazaar economies, where market monopolies are often based on codes of secrecy. Thus, brand economies can attract really large-scale external investment and carry out major predatory expansions. This transactional logic would apply equally to public institutions (such as universities, temples, or branches of government) and private companies or traders.

From the perspective of an archaeologist or economic historian, Fanselow's model has strong predictive value. And because it is grounded in the physical properties of commodities, it has the added advantage of being extendable to periods for which little or no written documentation survives. It is quite clear, for example, that major expansions in commercial networks can often be traced to precisely those historical junctures when standardised packaging is combined with specialised methods of quantifying, sealing, and otherwise marking goods. In each case, moreover, economic expansion is associated with an unprecedented concentration of capital, manifest in the appearance of centralised institutions such as temples and palaces, which stored and recorded information about the production and circulation of goods in new and more transparent ways (notably through the use of written media, in combination with complex systems of numerical notation).

The earliest, and in many ways paradigmatic, example of this historical pattern is the aforementioned Uruk Expansion of the 4th millennium BC, which lay foundations for similar commercial expansions in Egypt (Wengrow 2006), the Indus Valley (Miller 1985, 2008), Central Asia (Kohl 2007), and the eastern Mediterranean (see Bennett [2008] for an explicit discussion of palatial marketing in Bronze Age Greece). In his wide-ranging contribution to this volume ("Making and Marking Relationships: Bronze Age Brandings and Mediterranean Commodities"), Andrew Bevan uses archaeological evidence for four important commodity types (metals, textiles, oils, and wine) to offer a rich account of branding and marketing strategies in the 2nd millennium BC. His study provides a template for further archaeological work on these topics, showing how a close consideration of material culture patterning—combined with critical use of ancient textual and iconographic sources—casts light on forms of commodity fetishism that originated in the realm of elite culture, but also "leaked" into wider commercial arenas through both sanctioned and nonsanctioned processes of imitation. The implied parallels with modern celebrity culture are clear and intuitive, but Bevan is careful to emphasise the distinct contexts and media of cultural dissemination (both elite and nonelite) that prevent us from drawing quick and easy parallels between modern and premodern situations. We are left in no doubt, however, that Bronze Age temples and palaces, far from being mere collectors and distributors of economic surplus, also enlisted the support of the gods as promoters and manipulators of economic value (see also Sherratt and Sherratt 1991).

In considering a variety of commercial media, Bevan also develops a further argument of my original paper (Wengrow 2008: 9–11, 28; compare Miller 2008; Wilk 2008), which is that the intense fusion of mass-produced homogeneity and powerful cultural symbolism—an instantly recognisable feature of contemporary brands—is not unique to modern material culture. The archaeological record furnishes many examples other than those discussed above. Susan Sherratt (1999), for instance, finds similar qualities in the mass-produced assemblages of painted pottery that circulated throughout the eastern Mediterranean during the later 2nd millennium BC. She writes of the decoration of one such vessel-class (illustrated here in the chapter by Bevan as his Figure 4e):

> Again it is the multi-valency of the overall impression that succeeds in making these pots socially and culturally attractive: fine textiles, chariots, precious metal vessels associated with elite drinking rituals, bulls as symbols of divine or regal power in Near East and Aegean alike—all undeniably elite symbols, and all ... subjects of international exchange at the highest level; but all wrapped up in a run-of-the mill clay pot suitable for being lost forever in a tomb or for possession and use by someone who may belong to a less than super-elite stratum of society, but nevertheless has his social pretensions and aspirations. (p. 189)

In the local imitation and adaption by Cypriot intermediaries of established (Aegean) ceramic forms and decorative schema, Sherratt further identifies deliberate marketing strategies in the Bronze Age that are no different in their essentials from the contemporary phenomenon of fake branding. The double irony, surely, is that these ancient imitations not only command high prices on today's antiquities market, where they are sold as rare and authentic works of art, but also generate their own modern copies, fakes, and forgeries.

Brand Underworlds, Authenticity, and Informal Connoisseurs

The following chapter, by Magdalena Crăciun, presents a radical shift in the scale of analysis, taking us into the world of a modern Istanbulite *imitasyoncu*: a "maker and seller of imitations," whose skills in the manufacture and distribution of forged brands have earned him a stable position in the urban market. Yet many of the key issues raised in Bevan's archaeological appraisal of branding remain salient in this careful ethnographic study.

From the perspective of one man's choice of profession, Crăciun subtly problematises relationships between "formal" and "informal" structures of branding, showing how the latter, too, have their codes of morality and professional standards, their own networks of manufacturers and retailers. From this vantage point, she is able to question "the binary logic of the original and the counterfeit, and its built-in bias in favour of the former." Working constantly on the margins of the law, the activities of such individuals are as central to the construction of modern brand consciousness as those of more recognised actors such as professional advertisers or celebrities, even if they have learned, through hard commercial experience, "to disregard the myths with which brands are layered." Here, then, we gain detailed access to the social motivations and aspirations of an "anonymous contributor to the collaborative construction of a brand," whose work, "located in the underbelly of what is publicly acknowledged," nevertheless supports "the visible and the acknowledged" world of Armani, Calvin Klein, and Ralph Laurent.

In a complementary study, Rosana Pinheiro-Machado ("The Attribution of Authenticity to 'Real' and 'Fake' Branded Commodities in Brazil and China") explores the diverse social logics and moral imperatives that inform the consumption of what we might generically call "fake brands." Her case studies arise from an unusual comparison of China and Brazil, countries shown by statistical analyses to be the "greatest consumers of fake and pirated goods in the world." Contained within the study is also an exposition of the anthropologist's traditional focus on border zones, in this case the post-unification boundary between mainland China and Hong Kong, which also happens to be the location of one of the world's largest centres for commerce in fake brands. This chapter takes us far beyond a monolithic concept of "elite emulation" (a term also widely and sometimes indiscriminately employed by archaeologists) as an explanation for the consumption of fake goods. In a multi-layered analysis, authenticity (and its absence) is shown to reside—not so much in the commodities themselves or in legal systems of classification and quality control—but rather in the micro-processes of acquiring, displaying, and evaluating personal consumables. Lurking behind these snapshots of social life is an awareness of the transnational trade linking China and Brazil through the export of cheap manufactures, and hence of the interactions that flow between apparently disparate worlds of consumption.

Branding and the Techniques of the Body

Ferenc Hammer ("The Real One: Western Brands and Competing Notions of Authenticity in Socialist Hungary") offers an ethnographic window onto an unlikely phenomenon, which pushes our understanding of brand consumption into new territories. His chapter concerns the emergence, during the 1980s, of "a small niche of brand culture" in a Soviet Bloc planned economy, where formal marketing structures and retail outlets for Western goods were absent. This is an account of how foreign brands—specifically branded jeans—entered a situation where the charismatic labelling of clothing and consumables was a state monopoly and where imported brands were officially regarded as subversive, even dangerous influences on youth culture. Many of Hammer's informants recall how they "jumped into the bathtub to spend a crucial phase together with their jeans in which the latter would be shaped to the figure of the owner," a self-baptism that transformed branded jeans into a "second skin." Worn jeans often passed as gifts or inheritance between close kin and intimate friends, accruing personal embellishments and carefully managed signs of wear along the way. Here again, "authenticity" is seen to reside not simply in the branded object, but in a series of routines and performances that initiate the wearer of branded clothing into a new social and moral status. This chapter, like Alison Clarke's contribution (below), further illustrates how brand value can be retained, enhanced, transformed, or altogether exhausted through secondary and tertiary exchanges of goods, which fall outside the purview of conventional market research.

A very different perspective on branding and the construction of social identity is provided by Jean-Pierre Warnier's "Royal Branding and the Techniques of the Body, the Self, and Power in West Cameroon." Warnier's chapter returns us to the theme of branded goods in a world of kingship and sacred economies, this time observed through the eyes of the ethnographer rather than the archaeologist. A superficial examination of contemporary consumption in highland Cameroon might suggest that the introduction of such things as branded cosmetic lotions (and the "brand myths" projected by advertising) are currently transforming these traditional hierarchies (and their local ritual economies, based on substances such as palm oil and raffia wine) in accordance with the interests of external market logics. However, as Warnier demonstrates, the reality is quite the contrary. It is the external market that is being incorporated, as it has been for centuries, into an existing system of social relations that includes supernatural agents (kings, witches, ancestors) as well as human ones. Moreover, this overarching system is firmly anchored in what Marcel Mauss termed "techniques of the body," including traditional and effective ways of making, consuming, and exchanging things that are both highly durable and rarely articulated through speech or writing.

The absorption of imported goods, included branded products, in the Cameroon grassfields is further filtered through local concepts of trust and deception, customary means of evaluating any substance or object that comes into contact with the body and skin or that has the potential to give or take life (compare Burke 1996). Here, worries about the purity and reliability of branded commodities, especially those that come from outside the kingdom, are easily translated into concerns about the negative effects of witchcraft. For instance, to understand the significance of a mother routinely rubbing Pears Baby Lotion onto her infant's skin, it is first necessary to understand the ambivalent relationship between a sacred king and his subjects, as mediated through ritual protection of the body's surface and orifices. In developing observations such as these, Warnier exemplifies the depth of ethnographic observation required for a convincing analysis of how branded products are received and consumed across diverse cultural contexts. His account, bridging the exotic and the familiar, will no doubt be of wide interest to archaeologists and historians as well as anthropologists.

Branding and the World of Commodities

We remain in the realm of royal courts and sacred kingship for Jason Yaeger's archaeological study of "Commodities, Brands, and Village Economies in the Classic Maya Lowlands." As a setting for addressing the themes of this book, the social and economic landscape of the Classic Maya (350–900 AD) presents some instructive contrasts, as well as similarities, with the early Mediterranean economies discussed by Bevan. Particularly striking, by way of contrast, is the lack of mobility for bulk commodity transfers. Draft animals, wheeled vehicles, and sail boats were not features of the Classic Maya economy, and their absence is accompanied by an intriguing lack of evidence for such things as standardised commodity packaging and specialised sealing devices: the familiar material correlates of a "brand economy."

In Chapter 7, Yaeger orients his discussion around Fanselow's distinction between "brand economies" and "bazaar economies." Given that current reconstructions of Classic Maya economy are often highly polarised, with the very existence of marketplaces remaining a subject of contention, Yaeger is obliged to steer a careful course in applying such models to his main case study, from the Mopan Valley of western Belize. His conclusion, that close intimacy between producers and consumers precluded the development of either "bazaar" or "brand" economies, may not satisfy all. Nevertheless, this study begins to show how a focus on commodities and branding can provide stimulating alternatives to "thinking about the Maya economy as a Frankenstein hybrid of politically generative gift exchanges and functionally driven household provisioning."

Chapters 8 and 9 have a more "historical" focus, located between the remote past and the ethnographic present. Gracia Clark's "Lincoln Green and Real Dutch Java Prints: Cloth Selvedges as Brands in International

Trade" is a study in the kind of informal connoisseurship of fabrics that Bevan also detects for much earlier periods of history. In the absence of mass literacy, specialised advertising, and global communications media, Clark shows how the desires of African consumers have nevertheless shaped the production of specific textile brands in northern Europe over a period of centuries. Her careful study of manufactured variability in the appearance of cloth selvedge—the self-finished edge of a cut of cloth as it emerges from the loom—illustrates how messages regarding quality, provenance, and exclusivity can be subtly incorporated (literally woven) into more traditional media such as textiles, without necessary recourse to extraneous forms of labelling (and for a combination of the two strategies in the ancient textile trade between Mesopotamia and Turkey, see again Bevan this volume). There is a lesson here for archaeologists who would minimise the role of consumer demand in premodern economies, and equally for experts in contemporary marketing who would identify consumer manipulation of brand values as a recent trend, contingent upon technologies such as the Internet.

In "Of Marks, Prints, Pots, and *Becherovka*: Freemasons' Branding in Early Modern Europe?," Marcos Martinón-Torres presents an analysis of enigmatic marks, stamped on what is arguably the *leitmotif* of medieval commodity transformation and fakery: the alchemist's crucible. Resisting any easy interpretation of these marks as precursors to modern brands, Martinón-Torres explores the wider institutional background to alchemy in early modern Europe, highlighting intriguing (but ultimately ambiguous) links between the commercial dissemination of metallurgical equipment (as well as other crafted items) and the circulation of Masonic symbols. The resulting interpretation is clearly at odds with the traditional importance of secrecy among Freemasons and raises wider questions about the role of commercial institutions such as guilds in early modern Europe that seem best answered through a combination of textual and archaeological analysis (compare Gaimster and Gilchrist 2003). Martinón-Torres's chapter contributes to a growing body of scholarship on the history of European branding before the Industrial Revolution (Richardson 2008).

It seems appropriate that a comparative study of branding such as this should end where it began: within the household. In the concluding chapter, Alison Clarke ("The Second-Hand Brand: Liquid Assets and Borrowed Goods") shows how anthropological approaches can shed new light on what, for most readers, will be more familiar arenas of brand consumption than those described in earlier chapters. Her study focuses on the circulation of branded household consumables through eBay, now increasingly in the form of leasing arrangements rather than all-out purchase. Taking issue with contemporary marketing literature, and echoing points made by Pinheiro-Machado and others in this volume, she argues that the current fashion for "transient possession of brands" is not "simply the result of a rational economic consumer applying themselves to the newly rationalised territory of the post-primary market." Rather, the afterlife of brands within eBay's "perpetual archive of everyday material culture" also transforms their nature, "taming" them, and making them focal points for the dissemination of new cultural practices (e.g., of parenting, or commemoration) within the household.

By tracing the fortunes of branded products as used goods (things no longer "brand new"), Clarke's study also significantly complicates Fanselow's distinction between bazaar and brand economies. She shows us examples of branded goods circulating *within* a bazaar economy. But the on-line auction is a distinct kind of bazaar in which commercial identities are carefully tracked and transparently displayed by the awarding of points for trustworthiness and reliability. In this case, digital technology has helped create a hybrid arena for the circulation of commodities, suspended somewhere between Fanselow's ideal types of transactional order. Branded goods traded on-line have usually passed through the domestic sphere and might then be expected to have lost something of their "brand charisma." Yet Clarke demonstrates how, quite to the contrary, the "staging" of such goods on-line constitutes a form of reenchantment, drawing on the original brand but also adding new value through personal testimonies, experiences, and even surface marks acquired through use. Moreover, the seller's tastes (or rather a carefully edited version of them) are on show in the form of an inventory of branded products they have bought and sold. Consumers and buyers are able to become self-styled icons for particular brands, taking on something of the role of celebrities in first-phase brand consumption, and themselves "steering the pace of brand change" in the wider market.

Conclusion: Towards a Comparative History of "Decommoditisation"

Many of the chapters in this book argue for the primacy of social and historical context in the analysis of branding and branded goods. This should not, however, be taken as a plea for the hyper-relative. On the contrary, underlying all the contributions is an explicitly anthropological understanding of the phenomenon of branding, rooted in universal and perennial problems of human interaction. All, at some level, address a paradox of social life that rises to the surface whenever economy—broadly understood as the production, distribution, and consumption of material goods and immaterial knowledge—reaches a certain scale and level of uniformity. This is the paradox of sustaining personal relationships through the circulation of impersonal, standardised goods. We view branding as a strategy for resolving this paradox, grounded in the tension between two opposing tendencies of large-scale economies: the disenchantment of production to rationalise and maximise output and the reenchantment of homogeneous goods to make them acceptable for consumption and personalised exchange.

In its raw essentials, the social and material logic of branding—involving the detachment of goods from their unique contexts of production, their marking with central cultural symbols, and their subsequent reattachment to society in altered and elevated forms—bears comparison with the timeless and universal logic of ritual transformation, as discussed by anthropologists such as Maurice Bloch (1992). As I have argued elsewhere, and as exemplified by the chapters in this volume, what has varied over time and space is:

> the nexus of authenticity, quality control, and desire from which brand economies draw their authority; the web of agencies (real or imagined) through which homogeneous goods must be seen to pass in order to be consumed, be they the bodies of the ancestral dead, the gods, heads of state, secular business gurus, media celebrities, or that core fetish of post-modernity, the body of the sovereign consumer citizen in the act of self-fashioning. (Wengrow 2008: 21)

The broader conclusion is that contemporary branding strategies should not be studied merely as part of the long-term history of commoditisation, but also as part of its opposite: that is, the *history of decommoditisation*, on a continuum with techniques of gift-giving, ritual, and sacrifice, which have long been a focus of research in archaeology and anthropology. And it follows from this that a truly cross-cultural analysis of branding—of the kind recently called for by Schroeder (2005) and Cayla and Arnould (2008)—will involve a great deal more than just shifting the goalposts of consumer taste to include a wider variety of "brand myths."

References

Algaze, G. 1993. *The Uruk World System: The Dynamics of Expansion of Early Mesopotamian Civilization*. Chicago: University of Chicago Press.

Barthes, R. 1977 [1964]. Rhetoric of the image. In S. Heath (Trans.), *Image/music/text*: 32–51. New York: Hill and Wang.

Baudrillard, J. 1968. *Le Système des Objets*. Paris: Gallimard.

Baudrillard, J. 1970. *La Société de Consommation*. Paris: Gallimard.

Baudrillard, J. 1981. *For a Critique of the Political Economy of the Sign* (C. Levin, Trans.). St. Louis: Telos Press.

Benjamin, W. 1999 [1936]. *Illuminations*. London: Pimlico.

Bennet, D. J. 2008. Palace™: Speculations on palatial production in Mycenaean Greece with (some) reference to glass. In C. Jackson and E. E. Wager

(Eds.), *Vitreous Material in the Late Bronze Age Aegean: A Window to the East Mediterranean World*, Number 8 in Sheffield Studies in Aegean Archaeology: 151–72. Oxford: Oxbow.

Bloch, M. 1992. *Prey into Hunter. The Politics of Religious Experience.* Cambridge: Cambridge University Press.

Bourdieu, P. 1977. *Outline of a Theory of Practice.* Cambridge: Cambridge University Press.

Burke, T. 1996. *Lifebuoy Men, Lux Women. Commodification, Consumption, & Cleanliness in Modern Zimbabwe.* London: Leicester University Press.

Carrier, J. G. 1993. The rituals of Christmas giving. In D. Miller (Ed.), *Unwrapping Christmas*: 55–74. Oxford: Clarendon Press.

Carrier, J. G. 1995. *Gifts and Commodities. Exchange and Western Capitalism since 1700.* London: Routledge.

Cayla, J. and E. J. Arnould. 2008. A cultural approach to branding in the global marketplace. *Journal of International Marketing* 16:86–112.

Charvát, P. 1994. The seals and their functions in the Halaf- and Ubaid-cultures. In R.-B. Wartke (Ed.), *Handwerk und Technologie im Alten Orient: Ein Beitrag zur Geschichte der Technik im Altertum*: 9–16. Mainz, Germany: Philipp von Zabern.

Chevalier, M. and G. Mazzalovo. 2004. *Pro-Logo. Brands as a Factor of Progress.* New York: Palgrave Macmillan.

de Maaijer, R. 2001. Late third millennium identifying marks. In W. H. van Soldt (Ed.), *Studies Presented to Klaas R. Veenhof on the Occasion of His Sixty-Fifth Birthday*: 300–24. Leiden: Netherlands Institute for the Near East.

Eckhardt, G. M. and A. Bengtsson. 2009. A brief history of branding in China. *Journal of Macromarketing* (published on-line).

Fanselow, F. 1990. The bazaar economy or how bizarre is the bazaar really? *Man* 25:250–65.

Ferioli, P. Ed. 1994. *Archives before Writing*. Turin, Italy: Scriptorium.

Foxvog, D. A. 1995. Sumerian brands and branding irons. *Zeitschrift für Assyriologie und Vordasiatische Archäologie* 85:1–7.

Gaimster, D. and R. Gilchrist, Eds. 2003. *The Archaeology of the Reformation 1480–1580.* Leeds, UK: Maney.

Hamilton, G. G. 2006. *Commerce and Capitalism in Chinese Societies.* London: Routledge.

Hamilton, G. G. and C. Lai. 1989. Consumerism without capitalism: Consumption and brand names in late imperial China. In H. Rutz and B. Orlove (Eds.), *The Social Economy of Consumption*: 253–80. Lanham, MD: University Press of America.

Holt, D. 2002. Why do brands cause trouble? A dialectical theory of consumer culture and branding. *Journal of Consumer Research* 29:70–90.

Holt, D. 2004. *How Brands Become Icons: The Principles of Cultural Branding.* Boston: Harvard Business School Press.

Holt, D. 2006. Jack Daniels' America: Iconic brands as ideological parasites and proselytizers. *Journal of Consumer Culture* 6:355–77.

Holt, D. 2008. Response to D. Wengrow, "Prehistories of Commodity Branding." *Current Anthropology* 49:22.

Horkheimer, M. and T. W. Adorno. 1996 [1944]. *Dialectic of Enlightenment*. New York: Continuum.

Klein, N. 2000. *No Logo: Taking Aim at Brand Bullies*. London: Flamingo.

Kohl, P. 2007. *The Making of Eurasia*. Cambridge: Cambridge University Press.

Kopytoff, I. 1986. The cultural biography of things: Commoditization as process. In A. Appadurai (Ed.), *The Social Life of Things: Commodities in Cultural Perspective*: 64–91. Cambridge: Cambridge University Press.

Miller, D. 1985. Ideology and the Harappan civilisation. *Journal of Anthropological Archaeology* 4:34–71.

Miller, D. 2008. Response to D. Wengrow, "Prehistories of Commodity Branding." *Current Anthropology* 49:23.

Moore, K. and S. Reid. 2008. The birth of brand: 4000 years of branding. *Business History* 50:419–32.

Nissen, H., Damerow, P., and Englund, R. K. 1993. *Archaic Bookkeeping: Early Writing and Techniques of Economic Administration in the Ancient Near East.* Chicago: University of Chicago Press.

Pittman, H. 1995. Cylinder seals and scarabs in the ancient Near East. In J. M. Sasson et al. (Eds.), *Civilizations of the Ancient Near East*: 1589–1603. New York: Scribner.

Pittman, H. 2001. Mesopotamian interregional relations reflected through glyptic evidence in the Late Chalcolithic 1–5 periods. In M. Rothman (Ed.), *Uruk Mesopotamia and Its Neighbours*: 403–43. Oxford: James Currey/School of American Research Press.

Renfrew, C. 2001. Commodification and institution in group-oriented and individualising societies. In W. G. Runciman (Ed.), *The Origin of Human Social Institutions*: 93–117. London: British Academy.

Renfrew, C. 2005. Archaeology and commodification: The role of things in societal transformations. In W. van Binsbergen and P. Geschiere (Eds.), *Commodification: Things, Agency, and Identities: The Social Life of Things Revisited*: 85–98. Berlin: LIT.

Richardson, G. 2008. *Brand Names before the Industrial Revolution*. NBER Working Paper No. 13930. Cambridge, MA: National Bureau of Economic Research.

Roberts, K. 2004. *Lovemarks: The Future beyond Brands*. London: Powerhouse Cultural Entertainment Books.

Rothman, M., Ed. 2001. *Uruk Mesopotamia and Its Neighbours*. Oxford: James Currey/School of American Research Press.

Schroeder, J. 2005. The artist and the brand. *European Journal of Marketing* 39:1291–305.

Sherratt, A. G. 1999. Cash-crops before cash: Organic consumables and trade. In C. Gosden and J. G. Hather (Eds.), *The Prehistory of Food: Appetites for Change*: 13–34. London: Routledge.

Sherratt, A. G. 2004. Material resources, capital, and power: The coevolution of society and culture. In G. M. Feinman and L. M. Nicholas (Eds.),

Archaeological Perspectives on Political Economies: 79–104. Salt Lake City: University of Utah Press.

Sherratt, A. G. and E. S. Sherratt. 1991. From luxuries to commodities: The nature of Mediterranean Bronze Age trading systems. In N. Gale (Ed.), *Bronze Age Trade in the Mediterranean*: 351–86. Jonsered, Denmark: Paul Åström.

Sherratt, E. S. 1999. *E pur si muove*: Pots, markets, and values in the second-millennium Mediterranean. In J. P. Crielaard et al. (Eds.), *The Complex Past of Pottery: Production, Circulation, and Consumption of Mycenaean and Greek Pottery*: 163–211. Amsterdam: Gieben.

Tobin, J., Ed. 1992. *Re-Made in Japan: Everyday Life and Consumer Taste in a Changing Society*. New Haven, CT: Yale University Press.

Wengrow, D. 2006. *The Archaeology of Early Egypt. Social Transformations in North-East Africa, 10,000 to 2650* BC. Cambridge: Cambridge University Press.

Wengrow, D. 2008. Prehistories of commodity branding. *Current Anthropology* 49:7–34.

Wernick, A. 1991. *Promotional Culture: Advertising, Ideology, and Symbolic Expression*. London: Sage.

Wilk, R. 2008. Response to D. Wengrow, "Prehistories of Commodity Branding." *Current Anthropology* 49:25–26.

Williams, R. 1980 [1960]. Advertising: The magic system. In *Problems in Materialism and Culture: Selected Essays*: 170–95. London: Verson.

Winter, I. J. 2000. *Le palais imaginaire:* Scale and meaning in the iconography of Neo-Assyrian cylinder seals. In C. Uehlinger (Ed.), *Images as Media: Sources for the Cultural History of the Near East and Eastern Mediterranean (1st millennium* BCE*)*: 51–88. Fribourg, Germany: University Press Fribourg.

Winter, I. J. 2008. Response to D. Wengrow, "Prehistories of Commodity Branding." *Current Anthropology* 49:26.

CHAPTER 2

MAKING AND MARKING RELATIONSHIPS

BRONZE AGE BRANDINGS AND MEDITERRANEAN COMMODITIES

Andrew Bevan[1]

INTRODUCTION

This chapter begins from a position of broad agreement with the idea that modern Western brands are just a particularly extreme and influential example of a cultural practice that becomes necessary in almost any economy of a certain scale where there exists a combination of mass-produced goods, aspirational consumers, and transregional systems of exchange (Fanselow 1990; Foster 2005; Hamilton and Lai 1989; Wengrow 2008). In such cases, it becomes important to find ways to reinvest otherwise standardised and deracinated products with more singular social meanings. A key point that I will seek to make, however, is that this resocialisation is most commonly achieved via the same abstract models, cultural metaphors, and practical techniques that people also use to coordinate their social relationships with one another. Branding and related behaviours, therefore, deserve to be considered more explicitly through a body of sociological theory addressing how humans structure and cue for particular types of interpersonal relationship. The discussion below begins by reviewing how relational models fit into broader anthropological and marketing theory and then considers how object values and commodity brands relate to the operation of larger scale distributed economies. The case study that then follows explores the commercial entanglement and social roles of four defining commodity types in the Mediterranean world, from the Bronze Age onwards—metals, textiles, oils, and wine—and argues that we gain much insight into the conceptual metaphors of past societies, into modern academic debates, and into contemporary cultures of branding by approaching standardised goods from this relational perspective.

Theoretical Perspectives

Values, Brands, and Relationships

When we speak of the value of an object, we introduce a semantically ambivalent but socially powerful term. In English, for example, words such as "value," "worth," and "taste" all retain a curious charge due to the fact that they evoke both what we think of as a personally ascribed or natural property and what we assume is a socially negotiated one, what we seek to keep separate from our ethical life and what we instinctively associate with moral or immoral behaviour (Bevan 2007: 8–18). As Georg Simmel pointed out long ago (1900), value thus lies at the interface between what is perceived as objective and subjective, or, as Daniel Miller suggested more recently (2008: 1123), it is a term that people regularly use as a bridge to connect their market-led measurement of the world and a continued emphasis on the socially inalienable.

If theoretical perspectives on value have a much longer Western and non-Western intellectual history, there have been perhaps two particularly important academic contributions over the last couple of decades. The first is an emphasis on understanding regimes of value as they change over the course of an object's full and often quite complex career, during which that object can pass through a range of production, distribution, reinterpretation, and consumption states (e.g., Appadurai 1986). The second emphasis, far more patchily promoted, has been on pulling down many of the intellectual barricades separating, for example, the perspectives on value offered by Marxist theories of embodied labour, the utility functions of classical economics, Maussian gifts, and late 20th-century consumer theory (e.g., Aswani and Sheppard 2003; Goody 2006; McGraw et al. 2003; Wengrow 2008; also Bevan 2007: 8–25).

These developments leave us much better placed to adopt unashamedly synthetic approaches to object value and to place phenomena such as commodity branding into a broader cross-cultural and diachronic context. Branding is a label that has become a rather loose metaphor for a range of stylistic phenomena and persuasive agendas, but as a point of analytical departure, it is more useful to restrict the term to the realm of broadly substitutable goods circulating as traded commodities. There has been much research devoted to the process of commodification as a form of social alienation, but more recent work on commodities has repeatedly asserted how inherently social their subsequent promotion, reception and manipulation really is. My argument here is that the processes by which a personality is reattached (for the phrase, see Foster 2005: 11) to a standardised commodity often invoke the logics used to coordinate human social relationships. In fact, there is nothing particularly new in this assertion, and a whole strain of modern consumer research has emphasised the importance of "brand personality" and "relationship marketing," with respect to such varied topics as the rebranding of modern corporations as friends or family, the personification of branded objects, the genesis of conceptual "brand communities," the consumption strategies of recently immigrated groups, or the different standards that apply to different social classes when consuming fake brands (Aaker 1997; Aggarwal 2004; Muniz and O'Guinn 2001; Oswald 1999; and Pinheiro-Machado, this volume).

Branding in the modern Western world is often about establishing bonds of trust between producer, distributor, buyer, and seller (particularly the latter two). At least initially, substitutable branded commodities are a solution to the problem of navigating the very different levels of information that each of these groups might have about the finished product and therefore to the risk of being deceived about quality or quantity (what George Akerlof [1970] famously termed a fear of "adverse selection" or of buying "lemons"). The next section discusses how this solution relates to the particular configuration of exchange in larger scale, distributed economies, but for now my main point is simply that this fear of deception, and desire for trust, reflects a more fundamental problem underpinning any kind of interpersonal relationship: how two or more people establish predictable ways of behaving to one another as well as consensus about the ground rules that pertain to them in a given social context. Most of the interpersonal relationships in which humans engage are understood, organised, and signalled through a limited number of basic structural logics that encourage greater levels of social coordination and consensus. This has been a conclusion shared by a large number of different economic, ethnographic, and sociological commentators, but Alan Fiske has been particularly clear in his identification of four ways in which human beings habitually think through interpersonal relationships (1991; 2004b). Depending on social context, people can choose to emphasise (1) simple, undifferentiated relationships of inclusion or exclusion (what Fiske terms "communal sharing"), (2) ordered relationships of unequal status ("authority ranking"), (3) peer-to-peer associations ("equality matching"), or (4) certain flexible kinds of metrical relationship ("market pricing").

There are good reasons for treating such distinctions as slightly more robust than just another set of anthropological or sociological types. In formal terms, they reflect four fundamental scales at which we measure the real world (nominal, ordinal, interval, and ratio) and suggest not only the evolution of certain human cognitive proclivities, but also important variations in the way these proclivities are culturally implemented (Fiske 2000; also Haslam 2004). As a set of working distinctions, they are particularly attractive given (1) the common ground they share with a wider variety of existing sociological theory (e.g., some of the distinctions raised by Douglas, Mauss, Piaget, Ricoeur, Sahlins, and Weber amongst others: Whitehead 1993: 11–12); (2) their congruence with modern cognitive theories about childhood language development and the modularity of the human mind; and (3) the balance they offer between behavioural determinism and cultural relativism (Fiske and Haslam 2000).

If we turn to the role of material culture, people can use objects to advance their interpersonal relationships in the direction of any of these different logics: Gifts, for example, do not always reflect and promote the most famous case of reciprocal exchange but can otherwise involve fairly altruistic sharing within a community, ranked differences in social status, or market-led enticements (Komter 2001; also Bevan 2007: 25–26). Rather, it is the diversion of objects from one kind of socially agreed relationship with people to another that is usually the stuff of moral outrage (particularly the commodification of things in seemingly inappropriate ways; see Appadurai 1986: 14–16; Kopytoff 1986; McGraw and Tetlock 2005; McGraw et al. 2003; Tuk et al. 2009), and arguably the main reason for the semantic ambivalence of value/values in the first place.

More importantly perhaps, certain kinds of objects are much more likely to encourage certain kinds of social relationship than others, and material culture is one important way in which people's preferred relational logics are made physically manifest and socially constituted as norms (Fiske 2004a; indeed these physical manifestations are usually the ways that such patterns are recognised by children, outsiders, or anthropologists). For example, communal, categorical models for social relationships are often promoted by food-sharing paraphernalia, by emblematic body modification (and its representation on artefacts), by acts of physical intimacy (and associated objects), by initiation rites, and by purity laws or taboo. Ranked relational models are commonly established via physical props that choreograph social encounters in ordered, asymmetric ways (above/below, in front/behind, before/after, bigger/smaller, stronger/weaker). They can also be reinforced by the perceived natural hierarchy of material goods (e.g., Mauss and Durkheim 1963: 83–84). Peer-to-peer relationships of equivalence are usually forged and reinforced by objects that facilitate balanced, turn taking, or complementary contributions. Market-pricing logics tend to be encouraged by objects that facilitate easy convertibility and mensurability (e.g., those with bullion value), propositional offerings (e.g., samples), brand mnemonics, and accounting symbols (e.g., logos, labels, seals, and weights).

These physical propensities therefore suggest that object styles might be subject to both direct and indirect selection based on their suitability for certain kinds of social relationship, and hence that we might hope to identify the occasional residues of past relational models in the patterning of the material record. If we return to the case of modern Western commodity brands, some products make relatively straightforward, metrical claims about product efficacy (particularly those associated with a fairly early phase of advertising strategy; see Holt 2002: 80–81): For example, buy this soap and clean 50% more dishes. In other cases, the proposition is less precise but still metrical: Wear these clothes and have more friends. However, a distinctive feature of many branded commodities is that, while they are clearly created to streamline market-led relationships between people (by providing more standardised and recognisable quantities and qualities that alleviate fears of adverse selection), their brand image or brand personality often straddles one or more of Fiske's other three relational models (i.e., communal sharing, authority ranking, or equality matching), whether so engineered by marketing specialists or due to the creative contributions of consumers. This relational dexterity, often involving some kind of unspoken proposition, is a direct analogue to "indirect speech," where the use of veiled language avoids awkward mistakes in relationship coordination and/or can propose changes to existing relationships without incurring the full penalty of outright rejection (e.g., the language of bribes, half-spoken threats, and sexual come-ons; see Pinker et al. 2008).

Distributed Economies

Object value is clearly more malleable for some individuals, groups, and institutions than for others (Appadurai 1986: 31; Molm et al. 2001), and the overall social network within which such valuation occurs is an important structuring feature (e.g., Watts and Strogatz 1998). Many commentators have noted the importance of modern, hi-tech communication—cheap, fast, often literate, high-volume, long-distance—to what we think of as modern branding, including its ability to support new and often more ephemeral forms of entertainment, experience, community, and personal identity (Janson 2002; Muniz and O'Guinn 2001; Wallendorf 2001). However, while retaining an awareness that there is likely to be something unusual about the levels of branded material culture, branded spaces, and branded human identities present in the Western world of the last 50–100 years, we should really seek to place this in a much wider context with respect to information flow in a variety of different kinds of economy (e.g., Appadurai 1986: 48; Fanselow 1990; Hamilton and Lai 1989). The key issue is arguably the impact of drawn-out networks of production, exchange, and consumption. Indeed, although this used to be a less emphasised feature of modern brands, it has come rapidly back into fashion with an emphasis on "sourcing" by brand managers and anti-globalisation campaigners alike (Foster 2005: 11–12).

More broadly, we can associate these networks with the economies that have coalesced around sedentary, often highly urbanised societies in only a few parts of the world, particularly those that have developed trans-regional systems of exchange. In such circumstances, producers, distributors, and consumers become separated from one another both physically and culturally, and the propositional information that they might wish to exchange is navigated through a series of communication bottlenecks. While one-off luxury products occasionally retain elaborate biographies (real or imagined) as they pass through such bottlenecks, the same trick is rarely feasible for most goods.

The problems that this leads to are well known (e.g., Akerlof 1970; Richardson 2008): In many instances, consumers have far less knowledge about the potential range in quality of particular merchandise than the producers or distributors. Likewise, the conditions of sale or the nature of the object can often make it difficult to evaluate key properties such as durability, safety and effectiveness before purchase. Depending on their means, people may also not be able to afford to replace defective or otherwise poor purchases. This lack of economic flexibility tends to make them risk averse, but also often leaves them unable to test a seller's reliability through the trial and error of repeated purchase (though this depends on the goods involved). In many circumstances, there is also no easy or affordable legal recourse for those who have received defective goods. Finally, there are many incentives for producers and distributors to cut costs wherever possible.

In the face of these problems, commodity standardisation is an attractive solution, particularly if it is easily identifiable by the presence of some highly recognisable, carefully structured packets of cultural meaning. These mnemonic packets are usually created through physical addition to the objects involved (logos, labels, seals, special additives, and/or assembly practices), abstract symbolism, external advertisement, and, where possible, structured social performance, and thereafter are an efficient means of reinvesting standardised goods with more potent social identities (Foster 2005; Wengrow 2008).

In addition, we might also talk about three further interesting patterns associated with such systems. The first can be summarised by the English phrase "coals-to-Newcastle" (e.g., Fuller 1840: 542) and evokes the idea of trade in directions that seem senseless or uneconomical because of the superabundance of that particular commodity at the trade destination. Indeed, the Classical version of this phrase, "owls to Athens" (e.g., Aristophanes' *Birds* 301) is even neater because it refers, on the one hand, to the patron bird of the city, the associate of the goddess Athena, that roosted in the early Parthenon, and, on the other, probably also to slang for the abundant silver coins minted by Athens with a depiction the Athenian owl (hence, why take such owls to Athens where there were so many already?). In any case, a related, but structurally distinct phenomenon is the one in which an uneven geographic spread of technical knowledge, organisation, and labour that allows well-positioned intermediaries to add value to commodities in some manner (even if only through repackaging) and then pass them on to third parties or sometimes even back to the original source (e.g., Wengrow 2008: 11; Wilk 2006: 97). This is the foremost analytical feature of a "world-system," as originally formulated (Wallerstein 1974), in which very asymmetric economic relationships exist between core economic zones and their peripheries, based on a supra-territorial division of labour. Interestingly, there are plenty of examples of coals-to-Newcastle situations where these organisational asymmetries do not exist and where, instead, the key is in successful product differentiation (see below).

A second interesting feature of such economies is the charisma of standardised languages of practical efficiency and/or of elaborate consumer knowledge (Wengrow 2008: 8; see also what Michael Silverstein [2006: 493] calls "-onomic" knowledge). Luxury comestibles are a group of products for which this becomes particularly important: modern wine-talk, oil-talk, food-talk, and coffee-talk, for example, all have their own specialised terminologies, cultural authorities, and revered production practices. To varying degrees, they all share a semi-religious emphasis on provenance ("terroir," often with an emphasis on single estate products), timeliness (vintage, occasional production), the dialectic of technology and tradition (specialist equipment, secret recipes, authentic procedures), the long genealogy of the producer, and a specific jargon of production, distribution and consumption (Beverland 2005; Heath and Meneley 2007; Manning 2008). I will return to some of these features with respect to the marketing of Bronze Age oils and wine below.

A third interesting pattern associated with large-scale distributed economies can be loosely termed "leakage." While there are usually directional, carefully managed commodity flows in such economies and carefully drawn-up commercial and ideological agendas, there are also ways in which physical goods, skills, and brand abstractions escape from these channels into altogether different and often more informal contexts. This leakage is not an epiphenomenon, but a fundamental feature. For example, branded goods are often reinterpreted by consumers on their own terms, rather than the distributors. The reality of this practice and, for some, its ultimate desirability, is reflected in recent marketing initiatives that seek to influence only loosely the contexts in which goods are consumed, thereby leaving room for consumer innovations that might add further brand value (Arvidsson 2005: 243–44; Holt 2002). Likewise, branded goods are often recirculated as gifts or on the second-hand market in ways that feed back into their original value (A. Clarke, this volume). Sometimes this practice can overlap with the previous one in the physical reworking of branded objects to fit new consumer agendas (e.g., the intergenerational transfer of cut-off jeans; Hammer, this volume). Finally, brands often suffer from another kind of leakage when they are substituted for copies (see also both Crăciun and Pinheiro-Machado, this volume). The prepackaging and sealing practices associated with branded commodities are quality and quantity controls that reduce the risk of adverse selection, but they remain, like other efficient forms of communication (e.g., digital networks, genes) vulnerable to dissimulating free-riders: In a sense, imitation is a great indication of a (temporarily) successful brand.

A Protohistorical Perspective

Many of the theoretical approaches discussed above were developed with reference to modern Western practices or were derived from detailed anthropological fieldwork, with all of the rich levels of meaning that such research contexts provide. There is a danger, of course, that in turning to an archaeological case study, we become frustrated ethnographers (Shennan 2002: 9), attempting thick description but failing to substantiate it. The emphasis here, therefore, is placed on a proto-historic case—primarily the eastern Mediterranean in the 2nd millennium BC, but with a wider scope when necessary—that offers a rich combination of archaeological, textual, and iconographic information. This region and time period has been something of an intellectual battleground across which many of the major sociological, anthropological, and economic theories of the 20th century AD have been deployed in the hope of providing a decisive, ancient victory.

The urban growth and cultural expansion of 4th-millennium Mesopotamia was responsible for a pattern of city states, transregional linkages, and more standardised commodities that has had a profound and long-lasting impact on the history of the Middle East, Central Asia, Europe, and the Mediterranean (Wengrow 2008). The late 4th and 3rd millennia BC also see complex political, economic, and social structures emerging in the eastern Mediterranean alongside patterns of long-distance trade that became both more intensive and more unevenly felt with the increased use of donkeys for overland transport and the gradually expanding use of sailing ships. By the 2nd millennium BC (the main focus here), the eastern Mediterranean region (Figure 2.1) was broken up into a relatively well-defined set of polities. These took various shapes and sizes, but were typically under the direct rule of a king, and increasingly by the end of this period, under the indirect influence of a "Great King." Immediately below the royal family was usually a small and potentially factional, upper elite group that included extended family members and a range of other powerful individuals, typically from established aristocratic lineages. Many of these people enjoyed overlapping official roles as administrators, courtiers,

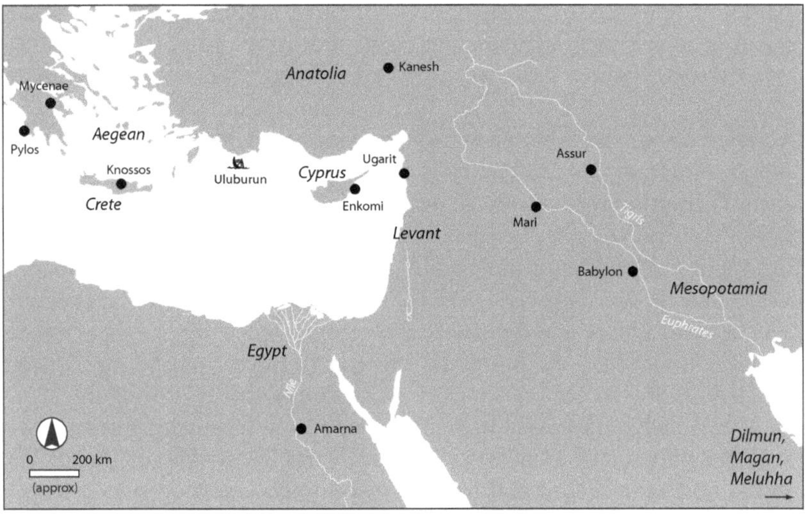

Figure 2.1. The eastern Mediterranean and Middle East with a list of places and regions mentioned in the text.

priests, traders, and patrons. Beneath this group in the social hierarchy there was usually a wider, lower elite group that held lesser bureaucratic posts and/or were less well-connected provincial figures. Beyond them was the rest of the urban and rural population whose archaeological and documentary visibility varies enormously.

In terms of exchange, we see a marked increase in the range and quantity of goods circulating across this area over the course the Bronze Age (roughly-speaking the 3rd and 2nd millennia BC: for an overview see Bevan 2007: 30–38; Sherratt and Sherratt 1991). Overall, we can document a pattern of expanding cultural and economic influence, westwards out of the core urbanised zones of Egypt and Mesopotamia and thereafter into the Levant, Anatolia, and the Aegean. Metals were a geographically restricted resource that were one of the driving factors behind long-distance exchange (e.g., Sherratt 1993), but textiles, oils, and wines were also important interregional trade goods, and, as we will see below, were commodities that travelled hundreds of kilometres despite the fact that they could, in principle, have been made anywhere in the region. These longer distance linkages encouraged an increasingly shared set of elite symbols. Counterfeits of popular commodities were very common, whether these were impressive synthetic versions such as "lapis lazuli of the kiln" rather than "lapis lazuli of the mountain" (i.e., blue glass instead of real lapis; see Oppenheim et al. 1970: 10–11) or merely cheap local copies. Indeed, imitations are present at all levels of the value hierarchy with, for example, plenty of evidence for copies of popular, but seemingly low-value pottery styles (also see below with respect to oil and wine).

One clear sign of this sharing of goods and ideas were weights and measures. From at least the mid-3rd millennium onwards in the eastern Mediterranean and Aegean, we see evidence for both regional systems of measurement and a familiarity with the points of convergence between them (Alberti and Parise 2005; Michailidou 1999; Pulak 2000; Rahmstorf 2006). These developments probably begin slightly earlier in Mesopotamia, during the 4th millennium, but in all cases, the evidence for standard weight systems co-appears with the evidence for more elaborate sealing practices and more standardised containers, suggesting, in each case, not only the up-scaling of administrative practice but also important changes in the relationship between society and material culture (Wengrow 2008).

By the later 2nd millennium in the eastern Mediterranean, there are even stronger patterns of metrical convergence and hybridisation, particularly in the northern Levant and northern Mesopotamia. Multiple systems of weights are found at the same site or on the same shipwreck, and the documentary evidence shows them together in everyday use by the same trader. Interoperable units of weight and capacity both facilitated and reflected a period of multi-language diplomacy, polyglot mobile communities, and deliberately cosmopolitan styles in a range of material culture, all of which encouraged neater, more syncretic packets of shared meaning in language, religion, and economic life. The major metric units in almost all areas of the eastern Mediterranean and Aegean were similar and often reference notional, practical measures such as a "donkey-load" or a "fleece." These allowed producers, distributors, and consumers to talk in counts of standard commodities rather than complex weighed measures and carefully described goods (see also Sherratt and Sherratt 1991: 362–63).

The language of both diplomacy and commerce had also become relatively uniform by the 2nd millennium BC (arguably, it was first established much earlier during the urban revolution in 4th-millennium Mesopotamia) and at its heart was a conceptual metaphor of the family and small village that was used to articulate far more complex relationships (Liverani 2000; Silver 1995: 50–53). Rulers referred to their entire kingdom as a royal household and estate (without implying control or ownership over it in a practical, everyday sense), and beyond this, carefully ranked their relations with other kingdoms, treating some as potential equals ("brothers"), others as vassals or superiors ("son," "father").

Likewise, commercial businesses were often organised around a real family, but wider partnerships with nonkin were very common, and the overall practice of commerce was choreographed through the language of colleagues-as-brothers, enticements-as-gifts, and firms-as-houses. This past conceptual framework that refers to more complex relationships in terms of the family and village has, ironically, wrought havoc with our interpretative models of Bronze Age economy and society. We have either conceived of this world as a kind of abstract, premodern "other" where temples and palaces were nodes of redistribution, objects moved in circuits of reciprocal gifting and private transactions were very limited, or have assumed that we can simply read off past economic activities using Western capitalist "common sense" about likely profit motives, object valuation, and business organisation. Perhaps the most famous version of this debate involves either agreement or disagreement with Karl Polanyi's suggestion (1957) that markets were largely absent from Bronze Age economies, but we can see similar academic fault lines running through much of the social sciences over the past 50 years.

Perhaps the best example of this, but by no means the only one, comes from the 14th century BC diplomatic archive found at Amarna in Egypt (Moran 1987). This set of letters includes correspondence between the Egyptian pharaoh and his clients in Egyptian-controlled areas from modern-day Gaza to Lebanon, and letters to supposed equals ("brothers") in states such as Assyria, Babylon, Hittite Anatolia, and Cyprus. Amongst the latter, there is great emphasis on communication through regular embassies and reciprocal greeting gifts, proper hospitality for visiting ambassadors, and occasional marriage alliances.

These interchanges have often been viewed as the ostentatious but distinctive workings of a classic premodern, Maussian economy. In fact, a wide range of studies on both the internal logic of the letters and other documents from the same period now make it clear that much else lies beneath the superficial phraseology (e.g., Liverani 1979, 2003: 123–25; Zaccagnini 1987; also Moran 1987: EA 39). For example, the bullion value of many of the gifts was of great concern, and occasionally less-experienced members amongst the corresponding kings slip into far more market-led discourse. There are counterintuitive, coals-to-Newcastle gifts (e.g., of ivory, ebony, copper, and gold to Egypt), however they may well be meant as explicit requests for the same commodity in return, but at levels commensurate with its greater abundance in the receiver's country. Many of the messengers and ambassadors between kings were also merchants. Some of the gifts were accepted in the spirit of brotherhood and equivalence, but then presented as tribute to an internal domestic population. Gifts may also have initiated a royal audience at which the disembarcation of other accompanying shipments could be discussed.

To revisit Fiske's relational models, these transactions were being coordinated via a logic of matched equivalence, but were also facilitating market-led calculations and were occasionally misrepresented as ranked differences in royal authority. Rulers were clearly aware of potential discrepancies between the phrasing of relationships and their reality. The king of Babylon alludes to this in his Egyptian correspondence when he expresses annoyance that his gift of chariots had been displayed as tribute during an Egyptian parade, and he tries to interfere in his rival's, the king of Assyria's, attempts to initiate relations with Egypt by deliberately drawing an ugly contrast between the two aspects of the process ("gifts" and "purchases"; see Zaccagnini 1987: 58). This veiled language of Bronze Age diplomacy successfully coordinated a range of political, legal, and economic relationships over very long distances, and greeting gifts were an established class of elite objects that were well designed to play flexible roles, as both real gifts between perceived equals of refined cultural sensibility, and bribes or commercial enticements.

I would also go further and argue that not only have we sometimes slightly misread the nature of market-led relationships in the Bronze Age Mediterranean and underestimated the overall amount of commercial activity, but we have also systematically misread the character of royal and/or temple disbursements. The balance of influence between palaces and temples varies significantly in different regions and at different times, with the palace being more of a constant throughout, but they were typically the largest producers, consumers, arbitrators, and patrons of Bronze Age commodities. Their invested wealth, organisational infrastructure, political authority, religious influence, control of taxation, and/or support for key port or caravan facilities often, therefore, configured the major patterns of commercial trade. However, this also made the ruler and the gods prime authenticators of commodity brand value. It is no accident that, for example, we see one or more gods as the celebrity guests for the wine festivals at Ugarit, as receivers of votive quantities of perfumed oil at Pylos, as the official guarantor of untampered textile loads at Assur, or as the statuesque overseers of copper production at Enkomi (see below). These were all goods that were primarily traded as marketed commodities, but divine or royal sponsorship was a crucial preliminary act, no doubt treated reverentially and sincerely, but with the longer term effect of endorsing the quality and quantity of the finished products.

In any event, the overarching conceptual metaphor of the family and small village binds together not only diplomacy and commerce, but also certain religious functions. The dwellings of rulers, gods, and traders had much in common semantically and practically. As suggested above, the term "house" had seemingly overlapping meanings that, depending on context, could refer to the household of an individual, a family-based firm, the house of a god (i.e., a temple), or the commercial interface of several of these (e.g., Castle 1992: 250–53; Killen 1979: 176–79; Silver 1995: 3–38; Veenhof 1972: 116, 397–99).

A further important consideration is the fact that even those commodities produced by the palaces and major temples for certain highly charged ceremonial events were not always consumed, permanently stored, or destroyed thereafter. Instead, they leaked out of this superficially closed loop into other circuits, and did so as a matter of course. An evocative example first mentioned by Leo Oppenheim (1964: 183–98) and recently revisited by David Wengrow (forthcoming) is the smoking meat of Egyptian and Mesopotamian ritual that was fed to divine statues incapable of eating it but thereafter reentered the world of circulating goods, imbued with enormous added charisma. John Bennet (2008) also discusses the semi-standardised, prestigious paraphernalia manufactured by the Mycenaean palaces for their elite supporters and notes that they involved materials so wholly infused with a palatial ideology and transformed by palatial craft specialists that we might construe them as a kind of Palace™, but were nonetheless objects that did eventually leak out into a wider, lower elite world.

The aftermath of ceremonial disbursement is just one instance of the kinds of brand leakage mentioned above as being widely relevant in the operation of distributed economies, and I will return to further examples and other types below. In any case, the choreography of commerce encouraged the resocialisation of commodities in particular ways, and we should therefore look particularly carefully at the way the physical appearance of such commodities, the marks made on them, and the labels attached to them might endorse not only quality and quantity, but also particular kinds of consumer relationship. The physical branding of cattle, sheep, objects, and slaves with the emblematic marks of temple, city, household, or individual ownership is attested from at least the late 3rd millennium BC in Mesopotamia if not before (de Maaijer 2001; Foxvog 1995).

Evidence for such marking practices occasionally survives in documentary form, but it is worth acknowledging the even greater problems associated with the patchy nature of the archaeological record. The most archaeologically enduring marks on commodities are often those made early on, during production, with those made during the later stages of distribution often being more ephemeral and fragile. Therefore, we are likely to get our strongest apparent evidence for product advertisement for those unusual cases where production and distribution are strongly integrated, such as for a limited range of royal and temple products. The patchy fate of labels and sealings is particularly important here: Clay sealings are only preserved under certain unusual burning conditions, and wax versions will have disappeared entirely. Ink inscriptions are sometimes found in better preserved contexts (particularly in Egypt; see, e.g., Hayes 1951), but their recovery is still incredibly limited and uneven. What the documentary and iconographic evidence does make clear, and what some of the specific examples discussed below also reveal, is that the wrapping and sealing of standardised goods was incredibly widespread and involved not just the treatment of individual items but also packaging in groups, often in very elaborate ways.

The following sections address the relationship between substitutable goods, product marking, and social relationships as they relate to the circulation of four key Mediterranean commodities.

Metals

Metals are by no means always the highest value items in early complex societies, but a conceptual triad of gold, silver, and bronze (the latter in the form of copper plus tin), is one that develops most clearly in this region from the 3rd millennium BC onwards (e.g., Sherratt and Sherratt 1991: fig. 2). Since the Bronze Age, a consistent and, to some extent, decisive feature of the Mediterranean economy has been the circulation of such metals in semi-standardised ingot form. There were a wide variety of different Bronze Age ingot shapes, including, buns, chains, rings, bars, and so-called oxhides, all showing some internal variation in weight but typically hovering around one known weight standard or another. This regularity meant that they could be assessed quickly by counting, even if they were often then checked by weighing and thereafter by remelting (see below, also Davies 1973: pl.lv; Dercksen 1996: 57–60).

Palaces and temples were clearly concerned with controlling a portion of the metals trade to safeguard their own production and, ultimately, also to set limits on the degree of access that other individuals might have to such prestige indicators. Some commentators see parts of the Bronze Age in terms of palatial monopolies in this regard, but typically this argument rests on palatial documentary evidence (that unsurprisingly reveals little interest in nonpalatial transactions). It is clear that the degree of palatial involvement did vary from region to region and over time, but there is nonetheless excellent evidence for the circulation of ingots, finished metal goods, and recycled scrap in private hands (Heltzer 1984; Zaccagnini 1984, and for a particularly enlightening Sumerian "dispute" between silver and copper, see Kramer 1963: 265). In any case, the key point I would like to establish below is that, while the restricted nature of metals as a resource led to some highly directional and sometimes carefully managed exchanges, the popularity of particular metal commodities, from particular sources and handled by particularly distributors, was something that had a lot to do with their marketing.

Most of the discussion below focuses on copper as a commodity because it is recovered archaeologically in greater quantities than other metals, has been subject to greater analytical scrutiny with regard to provenance, and was commonly used throughout the social hierarchy. However, it is first worth exploring the role of three other metals: gold, silver, and tin.

Gold comes from a limited number of sources in the eastern Mediterranean and Middle East and has further marketable physical properties: It is shiny, easily worked, and showy in the sense that it can be thinly spread over a large area as leaf and does not tarnish (e.g., Renfrew 1986). It is frequently, therefore, caught up in distinctions between mortality and transcendence, often understood as divine flesh, and used to coat cult statues and other objects (e.g., Aufrère 1991: 725–28; Wengrow forthcoming). Although there were gold sources exploited in Anatolia, the north Aegean, and further afield, much of the gold in the Bronze Age eastern Mediterranean and Middle East seems to have come from via Egypt, and Egyptian gold was sometimes further qualified by the names of the different mining districts from which it originated (Kassianidou and Knapp 2005: g.9.1; Lehrberger 1995: esp. g.1; Vercoutter 1959).

Silver was often a more solid, structural component of artefacts and cult images (in some instances, the bones of the gods; see Aufrère 1991: 412–13), with sources in Anatolia, the Aegean and via the Persian/Arabian Gulf (Moorey 1994: 219–20; Wagner et al. 1985). In addition, although gold, silver, and copper were all sometimes used as notional equivalencies for exchange purposes, silver establishes itself as by far the most common referent from the 3rd millennium onwards (Foster 1977: 35; Leemans 1960: 130–31; Powell 1999).

Compared to gold and silver, tin was an even more geographically restricted commodity and, although there may have been some limited (and perhaps early) exploitation of south-central Anatolian and/or central European sources (Pernicka et al. 2003: 160–65; Yener et al. 1993), a good deal of the tin used in the Bronze Age eastern Mediterranean seems to have been brought thousands of kilometres from the mountains of central Asia (Cierny and Weisgerber 2003), along a series of highly attenuated overland, riverine, and maritime routes. We have very little evidence for the early stages of this journey, but documentary and archaeological evidence becomes more available once this commodity reaches the Mesopotamian region.

One set of documents that have proven particularly insightful have been those pertaining to an early 2nd-millennium donkey caravan route from Assur across northern Mesopotamia to Kanesh in central Anatolia (see Figure 2.1). This Assyrian trade is discussed in greater detail in the section on textiles below, but the documentary archives from Kanesh and other sites offer clear evidence for the steadily increasing price of tin (in silver) as it travelled westwards (Joannes 1991; Veenhof 2003: 115–16; Veenhof and Eidem 2008: 82–83).

Beyond this geographically afforded overall trend, local variations in the exchange rate of tin were common, fairly sizeable, and the source of much commercial speculation. Some of these related to the varying local supply of other metals and other commodities involved in local trades, but politics was a major factor. For example, the kingdom of Mari temporarily gained much better access to tin supplies after a shift in the political of power among its eastern neighbours (including the demise of Assur) and, for a period of a few years in the 18th century BC, was able to buy tin at about half the usual price and thereby set up a whole tournament of Syrian regional diplomacy, as other states jostled for preferred local trading rights (Charpin and Durand 1991; Limet 1985: 16–17).

Tin was primarily valuable in the Bronze Age for alloying with copper to make bronze, which was harder and more durable. Overall, the copper trade was on an entirely different scale to that of gold, silver, and tin, therefore providing us with a much wider scope for analysis. The link between copper and commerce is quite strong: It was the major material used for tools and was commonly traded, recycled, and alloyed. Merchants on the Nile are referred to as being "as busy as copper" (Blackman and Peet 1925: 288), and copper was also a commonly used unit of equivalence (particularly in Egypt; see Janssen 1975: 441–42). A complex range of distinctions were made to describe different grades, types, and sources (Dercksen 1996: 33–47; Moran 1987: EA 33.9–18, EA 40.6–15; Pritchard 1969: 356), that referred, for example, to the copper's purity and colour, shape as an ingot, and whether it was whole or broken up. In the case of the Old Assyrian trade, the finer grades of Anatolian copper were sometimes worth over twice the price in silver of the poorer ones.

Indications of provenance were also particularly important and could involve references to broad regions, mining, areas or distribution points. Copper can be found in a wide variety of different places across the eastern Mediterranean and Middle East, and smaller quantities continued to come from diverse sources throughout the Bronze Age. However, to judge from both the archaeological and documentary evidence, two regions became dominant transregional suppliers: Dilmun/Magan and Alasia. Dilmun refers to the area of modern day Bahrain, whose population was well known for trading in metal and other resources coming up the Persian/Arabian Gulf from Magan (the metal-bearing zone in the al-Hajjar mountains of modern-day Oman) and via Meluhha (the Indus Valley). In other words, Dilmun was an intermediary rather than a primary metal producer, but nonetheless, the documentary sources often adopt the term "Dilmun copper" and suggest its trade was particularly important during the late 3rd and earlier 2nd millennia BC (Leemans 1960: 121–23; Weeks 2003). A second major region associated with copper production area was Alasia (from the Troodos mountains of modern-day Cyprus; for a recent discussion of the debate over the location of this place name and the petrographic provenance of the letters from the king of Alasia, see Goren et al. 2003). There are textual references to suggest it begins to become important during the earlier 2nd millennium, although at present this is only obvious archaeologically by the later Bronze Age (Muhly 1996: 49).

Amongst the range of Bronze Age ingot types, perhaps the best known and most archaeologically obvious is the oxhide ingot (Figure 2.2c). This shape was used for a variety of different metals, but was particularly associated with copper and, from at least the 15th to the 12th century BC; it appears on artefacts, in wall-paintings, and as a discrete sign in at least one contemporary script (e.g., Linear B). Literally hundreds of actual examples have also been recovered archaeologically over a very wide area of the Mediterranean and beyond, at least as far north as Bavaria and Bulgaria, as far south as Egypt, west to Sardinia, and east to the Euphrates (Gale 1991: g.2; Pulak 1997: 234–35). The weight of these ingots does vary (both in real terms and due to subsequent corrosion), but hovers around the range associated with various eastern Mediterranean talent-weight standards (*ca.* 27–30 kg; see Pulak 2000: 141–43), and the written documents confirm that this often allowed the ingots to be referred to in counts, with greater precision thereafter achieved through weighing.

This was the right weight for one side of a donkey or that an individual could carry. It was also a shape that could be stacked in a series of overlapping and stable rows in the hold of the ship (e.g., as found on the Uluburun wreck; see Pulak 1998: figs. 4, 12). "Oxhide ingot" is a modern label and it remains far from clear whether this symbolic connection also existed in the Bronze Age, but the shape does resemble traditional, semi-standard units of value in the form of animal skins (e.g., the oxhide and the fleece), is occasionally shown in association with live cattle or animal skins (e.g., Tomb of Ramses III: left wall upper; for cattle as another Cypriot export, see Hellbing 1979: 80), or handled in a manner akin to livestock and skins (thrown over the shoulders like a young lamb or calf, underfoot like a floor-covering, or shot at as if hunted, see below).

Lead isotope analysis most commonly sources the copper in many of the ingots to the Cypriot Troodos mountains (e.g., Stos-Gale et al. 1997) and by the later Bronze Age, we see copper oxhide ingots connected with some of the clearest Cypriot religious imagery as well as with other distinctively Cypriot manufactured goods (see below). However, the key point that the discussion below seeks to make is that this trade in Cypriot metal should not be interpreted simply as the export of raw material from a geographically favoured locale but as something that required regular and intensive promotion. There is no inherent reason, for example, why a state such as Egypt should have sought large quantities of copper from Cyprus, given the substantial supplies it had in the Sinai and Eastern Desert. Rather, a crucial factor was the positioning of distinctive commodities (a better way to construe not just elaborate finished metalwork but

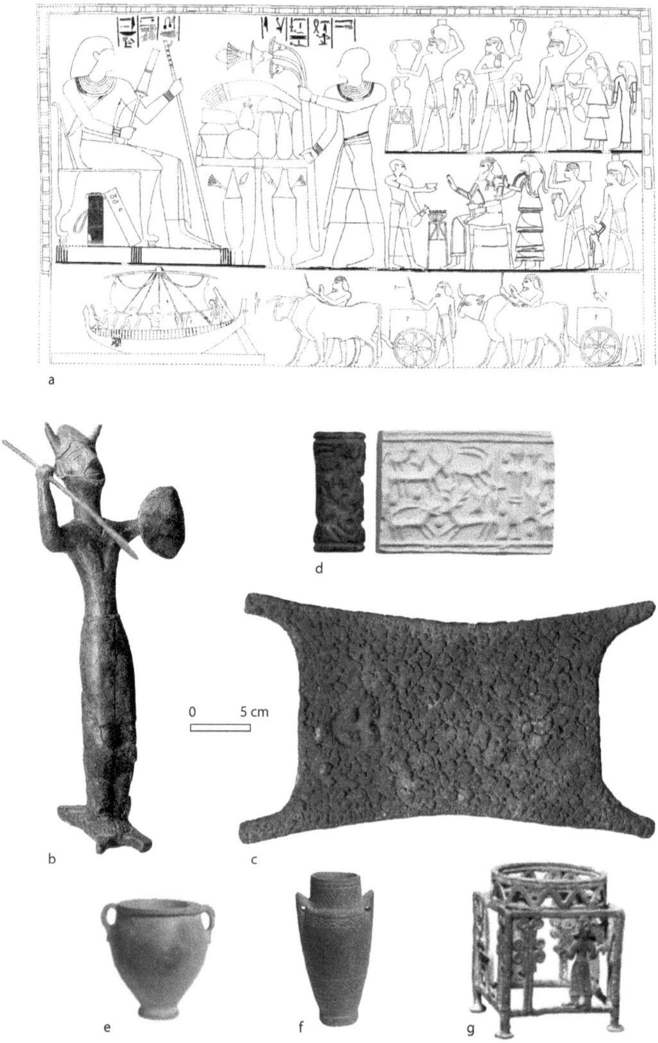

Figure 2.2. Copper and the symbolism of associated commodities: (a) a depiction from the tomb of Nebamun (after Säve-Söderbergh 1957: 25–27, pl. xxiii, courtesy of Oxford University Press); (b) a statuette of a (probable) god standing on an oxhide ingot (courtesy of the Department of Antiquities Cyprus); (c) a copper oxide ingot with an impressed stamp (shown at half of the stated scale); (d) a chloritite vessel imitating a metal version; (e) a chloritite vessel with incised decoration; (f) a cylinder seal showing a human figure with spear, oxhide ingots, a bird, a deer, a goat, a dog, an ox, and possible bun/ring ingots (shown at twice the stated scale); and (g) a bronze stand with a figure holding an oxhide ingot. All images not otherwise attributed are courtesy of the British Museum.

also the ingots themselves) and the sometimes awkward knitting together of upland landscapes, lowland entrepots, and overseas consumers.

The purity of these ingots (often 99% copper) implies not just primary smelting, but also a second stage of refinement (Hauptmann et al. 2002; Merkel 1986). However, judging the quality of metal commodities by eye (e.g., purity, internal flaws, etc.) has always been a difficult task (Richardson 2008: 8–11), and issues of verification were clearly a Bronze Age concern. For example, the Babylonian king in one Amarna letter complained about the quality of gold sent to him from Egypt (presumably after he has remelted it) and queried whether it had been inspected prior to departure (Moran 1987: EA 10).

Whole ingots were therefore not always instant indicators of quality and, in fact, freshly broken ones were in many ways more attractive, because they allowed limited inspection of the inside (e.g., Dercksen 1996: 58–59). In this regard, it was not just important that the oxhide ingot was convenient to handle, a visually familiar symbol of value, and of a semi-standard weight, but also that it bore more detailed technological and/or procedural trademarks. The first of these relates to the way the ingots were cast, in an open mould that left the upper side with a characteristic blistered surface (Hauptmann et al. 2002: 4; what would be the "hairy" side if it was, in fact, meant to evoke a real oxhide). This method, combined with the residual impurities in the metal, also meant that the ingots were sufficiently brittle that they could be broken up into smaller units. Given the need to signal that all of these qualities were present in the finished metal commodity, it is no accident that many of the representations of ingots in Bronze Age iconography do not merely depict the shape, but are also careful to show this stippling on one side (e.g., Davies 1973: 21 n.28; Karageorghis and Papasavvas 2003: figs. 1, 3).

A second set of reassuring marks were those cast, impressed or incised on perhaps half of the archaeologically surviving ingots (e.g., Figure 2.2c). As George Bass pointed out (1967: 73), these small marks were value-laden features commonly found on ingots from this point onwards, right through the Classical, Roman, and Medieval periods and up to the present day. In many later instances, they guarantee a certain metal purity, weight, provenance, and/or treatment en route. However, given that not all Bronze Age ingots are marked in any particular archaeological context, it seems more plausible that only one or more amongst a larger batch were being treated in this way. Likewise, different marks were clearly applied at different stages of manufacture and transport (not unlike the distributed marking practices associated with oil and wine containers, see below). Some were designs made in the mould itself or were impressed into the metal while it was cooling, and must therefore be associated with those doing the final refining and casting.

In contrast, the large number of incised marks probably reflects the concerns of the distributor, to identify the source, destination, and/or batchload, or in some limited instances as a temporary tallying mechanism. Marks depicting maritime images such as boats, fish, fishhooks, and tridents (e.g., Pulak 1998: g.10) are quite common and reemphasise (whether deliberately or accidentally) their product's circulation via maritime exchange. What all of this suggests is that we cannot treat Bronze Age ingot marks as immediately analogous to later hallmarks, for example, but instead should arguably see them as just one more conspicuous characteristic of proper trading practice that had both a practical role and a reassuring quality for the end-user.

Casting techniques and ingot marks lead us firmly to the issue of how metals production and trade was organised and who eventually had access to such commodities. Further evidence for the importance of distributors in the metals trade is suggested by the fact that the names associated with metals are often intermediary regions or coastal entrepots rather than the metal-producing areas themselves, hence "tin of Meluhha" or "Dilmun copper" (for a similar situation with Medieval "Damascus steel," see Feuerbach 2006). Likewise, the only known archaeological example of an oxhide ingot mould comes from a coastal entrepot some distance away from any metal-bearing zone (Lagarce et al. 1983: g.15) and the Cypriot sites with the greatest connection to ingot production, rather than primary smelting, are also lowland ports (e.g., Enkomi). The metals trade is also clearly an area of economic activity in which palaces (and temples) are very interested. For example, in a range of documentary sources, we see royal negotiations for shipments of gold, silver, and tin, but also much larger quantities of copper (sometimes involving several tons in one transaction). In various palatial archives, we can also document small to medium-sized disbursements of metal to smiths and careful attention to recovering the value of these as finished objects (e.g., Heltzer 1982: 91–95; Ventris and Chadwick 1973: 509; Wiseman 1953: 105–06). This pattern of involvement with metals has led many commentators to suggest that most new metal passed through the palace economy or leaked out into other circuits indirectly through recycling or tomb robbing (Weeks 2003: 140–42; see also Michailidou 2001; Sherratt 2000).

However, even leaving aside the eloquent evidence from the Assyrian trade, there is plenty of indication from elsewhere of noninstitutional trades and private ownership of metal bullion (Heltzer 1984; Zaccagnini 1984). A good example from Egypt comes from the late 15th century BC Theban tomb of Nebamun who was a physician and receives an oxhide ingot from a Syrian man, presumably in part payment for his professional help (Figure 2.2a; Wachsmann 1998: 46). A related question, however, is the presence or absence of self-organising, guild-like structures amongst the merchants and craft specialists of this region in the Bronze Age. In a variety of documentary contexts, we have lists of individuals in their professional groupings (and occasionally with their own leaders), but it remains very unclear whether these were just a bureaucratic convenience or whether they actively promoted product quality and identity in the manner that some Medieval guilds seem to have done (e.g., Cutler and Macdonald 1977; Gordon 1956; Ventris and Chadwick 1973: 509; see also Martinon-Torres, this volume; Richardson 2008).

In any case, the Nebamun example is also interesting because, alongside the ingot, both a large Canaanite-style transport jar and a pottery juglet (the latter probably containing perfumed oil) are shown as part of the overall transaction. Similarly, in a letter to his Egyptian counterpart, the Cypriot king requests payment, in large sums of silver, for a shipment of both copper and timber (Liverani 1979: 100–01; Moran 1987: EA 35.27–9). Elsewhere I have suggested that some of these wider, mixed exchanges involving copper implicitly document the intensified commodification of a whole upland, ophiolite landscape (also Bevan 2007: 174–79). The broad range of metals, stones, and high biodiversity present within or adjacent to such landscapes made them important sources for a whole range of products in addition to metals, including ultramafic softstones, timber, wildfowl, aromatic plants, and oils and good opportunities for grazing (hence also wool, meat, dairy, and horses). Some of the products were heavily interwoven in terms of their practical uses: Cypriot timber was a natural accompaniment to copper as it provided transport (ships), fuel for metallurgy, and perhaps structural support in mines. Likewise, ultramafic softstones (steatite, chloritite, serpentinite) were variously useful in the metal industry as heat-resistant materials for tuyères, moulds, and/or as finely-carveable stone for seals and vessels (Figure 2.2d–f).

So, a wide array of products was flowing with the metal out of the Cypriot upland zone and through major coastal entrepots. One interesting aspect of this process is that this mixed assemblage of commodities takes on many consistent symbolic associations, which both acted as propositional devices in a commercial context and as a way of articulating the tensions inherent in such upland/lowland, distributed metal economies. The first point to note is divine sponsorship of Cypriot copper production as articulated at lowland centres. Perhaps the most famous examples are bronze statues of (apparently divine) figures standing on oxhide ingots (Figure 2.2b), but oxhide ingots also appear as miniature votive items (and/or very small bullion units) and there are clear architectural associations between cult activity and copper processing areas at several Cypriot coastal sites (Knapp 1986).

Beyond these immediately religious associations, oxhide ingots had a wider symbolic capital as targets for Egyptian royal archery contests, as badges of personal wealth and overseas knowledge on Egyptian tomb walls, and as iconic devices shown on other Cypriot craft products such as bronze stands and softstone seals (Figure 2.2d and g; Davies 1935: 49–51, g.4). This contextual promiscuity reflects the proliferation of a highly successful brand sign and its interweaving with a host of related commodities and international activities. In this regard, both the stands and sealstones are worth brief further mention. The former were four sided, often wheeled stands for supporting large metal vessels (Figure 2.2g), and products of advanced metallurgical know-how; several have images of people carrying oxhide ingots (Figure 2.2g; Karageorghis and Papasavvas 2003). Such stands, and at the end of the Bronze Age perhaps iron daggers and knives (see Sherratt 1994), were desirable novelties in their own right that were traded across the eastern and central Mediterranean, but they were also promotional devices that reinforced notions of Cypriot innovation in the metal industry.

Specific types of Cypriot sealstone also show oxhide ingots, in association with plants, animals, and people all plausibly drawn from the same upland landscape in the Troodos mountains of Cyprus (Figure 2.2d; Graziadio 2003; Webb 2002: 118–26, pl.iii.1–8). The material of the seals (chloritite) almost certainly comes from this same general region and was heavily used in the metal industry (see above). As David Wengrow has suggested (2008: 8), seals are well designed to play a dual role "as components of bureaucratic systems and as charismatic signifiers of product identity," and these particular examples record, evoke, and promote the broad range of commodities flowing out of such upland landscapes.

The interesting thing about these distributed economic relationships and extended ophiolite brandings is that they are by no means unique to this region. If we jump backwards in time to the late 3rd millennium BC and across to the Persian/Arabian Gulf, a strikingly analogous case is provided by the documented flow of copper, timber, and other commodities down the Gulf towards the large urban centres of lowland Mesopotamia (Leemans 1960: 121–27; Weeks 2003).[2] Some of the products were coming from as far afield as the Indus Valley, but many were produced in a region known as Magan (particularly the ophiolite zone of what is now Oman), and were often shipped by better placed, intermediary traders from Dilmun (from around modern-day Bahrain). In a similar manner to Cyprus, intensified copper trading seems to have stimulated production of a range of ultramafic softstone products, including vessels, talc-tempered pottery, and moulds, but also the sealstones that Dilmun traders often used as part of their business (Bevan 2007: 175–77, especially fig. 8.21). In both the Cypriot and Dilmun/Magan cases, the symbolism of these ophiolite products expresses a certain *Metallschock*, oscillating between images of tradition and technical innovation and reflecting a socioeconomic relationship between upland communities, lowland distributors, and overseas consumers that was sometimes profitable and sometimes traumatic (for other examples of such asymmetry and its consequences, see Shennan 1999).

Textiles

Reading Gracia Clark's description of Medieval and modern cloth branding practices (this volume) in parallel with Klaas Veenhof's analysis (1972) of textile vocabulary in the Old Assyrian trading colony archives is, in many ways, to visit parallel worlds. In both cases, standardised bolts of cloth (enough to make one if not more full garments) are handled by wholesalers who cater for very specific local tastes. Groups of producers make distinctive products conforming to categories that outside buyers can identify. A clear and widely agreed set of types and quality grades are present and associated with many of these, also indications of provenance (some still meaningful, other long since become notional). Marks on the edges of bolts of cloth, on finished garments, and/or on packets of several textiles are important signifiers of reliable value.

To take the 2nd millennium BC evidence in greater detail (for what follows, see in particular Larsen 1987; Veenhof 1972, 2003), the Assyrian traders mentioned above were shipping tin and textiles westwards from Assur to central Anatolia (see Figure 2.1), sometimes indulging in local Anatolian trade in copper and other items, but ultimately looking to take their profits back to Assur in the form of silver and gold (which were lightweight and highly convertible). These mixed caravans of tin and textiles make an interesting combination. As suggested above, tin was a geographically restricted resource making its way to Anatolia (and beyond) along a trade route thousands of kilometres long. In contrast, the mechanical technology, basic knowhow and raw materials required to produce textiles were common across much of the Mediterranean and Middle East. For textiles, what therefore mattered was the ability of producers and traders to market products in standardised sizes and qualities, with recognised types of wool, distinctive manufacturing techniques, and/or popular types of decoration. In fact, the textiles in the Assyrian caravans heading to Anatolia included many non-Assyrian products that were imported from Babylonia to the south or acquired from nodal communities along the route, but also made locally at Assur.

There were a range of private individuals, partnerships, and institutions (e.g., the ruler of Assur, various temples) who invested in this trade, which involved the movement of large quantities and significant financial returns (conservatively, thousands of textiles, and hundreds of kilograms of profit in silver per year; see Larsen 1987: 51; Veenhof 2003: 70; Veenhof and Eidem 2008: 90). Our documentary sample is, of course, partial, and the texts focus very much on logistical arrangements rather than either the local production or initial purchase of the textiles in Assur or their final sale. Consignments arriving in Anatolia, for example, were entrusted as credit to other traders who were then responsible for marketing them locally. However, while we lack a clear view of the consumer end of the process, there are more than enough clues in the complex choreography of wrapping and sealing these items for transport to suggest that a kind of brand value was very important.

Both tin and textiles were packaged individually and/or in groups, using a standard cloth wrapping and according to standard approximate weights. Packets of textiles of around 30 kg each were sealed with hemispherical lumps of clay ("bullae"; see, e.g., Özgüç and Tunca 2001: 135) and placed on either side of the donkey. Other, smaller quantities were stored in top-packs that could be traded or added to en route. The clay lumps were sometimes stamped with the seal of the god and city of Assur as an important guarantee of officially verified quality, type, and quantity (related also to customs tax procedures), but most seals and other marks appear to represent particular individuals and/or their family-firms (Larsen 1977; Veenhof and Eidem 2008: 114–17). The making and breaking of sealings was generally only done at certain times (e.g., not before the end of the journey from Assur) and in front of witnesses: One of its primary roles was to endorse reliable levels of quality, quantity, and type. The proxy use of seals by third parties to stand for physically absent individuals and their family firms also suggests an important role for these as market abstractions, meant to propagate the reliability of a particular firm or of the Assur city trade in general.

This wrapping, marking, and sealing of textiles also occurred at several different levels, with certain procedures associated with textile bundles and others with individual pieces; with some marks or seals to do with the responsibilities of the trader and others to do with the owner and/or producer (Veenhof 1972: 41–44). However, for the individual pieces, a critical zone of attention seems to have been the textile border (*sissiktum*). In the early 2nd millennium BC, this area of a textile had great metonymic value: For example, the *sissiktum* was used as an emblem of the person themselves for divorce rituals, as a stand-in for an absent individual in certain haruspical rites, or in lieu of a readily available seal for everyday sealing purposes (Durand 1988: 40; Larsen 1977: 98; van der Toorn 1996: 46–47). Given that the Assyrian caravan trade was often involved in transporting bolts of uncut woven fabric (Veenhof 1972: 89–97), it seems sensible to assume that the term *sissiktum* could refer to the borders of both finished garments and uncut cloth.[3] In any case, these were the parts of wrapped textiles that were most visible to the buyer and therefore a key locus for branding activity (G. Clark, this volume).

What sorts of quality differences were such marking practices meant to endorse? The sheer variety of textile terms in the archives attests to a vast array of distinctions based on colour, provenance, finishing, style, and thickness (Veenhof 1972: 144–213). At the very top of the scale, were textiles of so-called royal quality that are referred to not only in the Old Assyrian caravan trade, but also in a range of other Bronze Age textile industries (see below). In some instances, this may literally mean "of or for the king," and thereby refer to royal trade products or those that had somehow found their way into general circulation, but sometimes it just seems to refer to the very finest quality garments. This is a good example of how Bronze Age royal brands might leak out to a wider elite group and will be revisited below with respect to the marketing of high-quality oils.

In any case, provenance tags were another means of marking out different styles of products in the caravan trade, and thereafter different perceived quality grades. For example, there were clear separations of different Anatolian, Assyrian, and Babylonian-style products. More precisely, the particular case of "Abarnian" textiles is interesting because the name implies a provenance from the town of Abarna, but the texts include at least one reference to these being made by a local Assyrian woman at Assur (Veenhof 1972: 123, 156–58, 191). The Medieval evidence for cloth making is particularly eloquent on such issues (G. Clark, this volume; Richardson 2008: 21–22): The main producing and distributing towns often gave their names to famous products and had a predatory brand quality, swallowing up the products of nearby places, and a key indicator of successful town-brands has been their survival into modern language (e.g., "worsted," "muslin," etc.).

In any case, the import substitution suggested by the Assur-made Abarnian textile is typical of well-placed distributors (see also below for oils), but conversely also a concern for those looking to uphold the value of existing imports. Another good example is a decision by Assur's commercial leaders to put a halt to any trade by expatriate Assyrians in two specific types of local Anatolian textile. This verdict was accompanied by the threat of heavy fines, seemingly because of a fear that the local products were either being marketed more fiercely by Assyrian traders than the actual imports or that they could be converted into cheap imitations (Veenhof 2003: 89–90; and compare with Crăciun, this volume).

The above discussion should make it clear that a whole range of product marking and market-led manipulation was behind the perceived value of these textiles, with good evidence for the kinds of symbolic abstraction that we commonly associate with modern branding practices. While the focus here has necessarily been on the richest body of documentary evidence (given the poor archaeological preservation of textiles), there are signs that such practices were widespread in the Bronze Age eastern

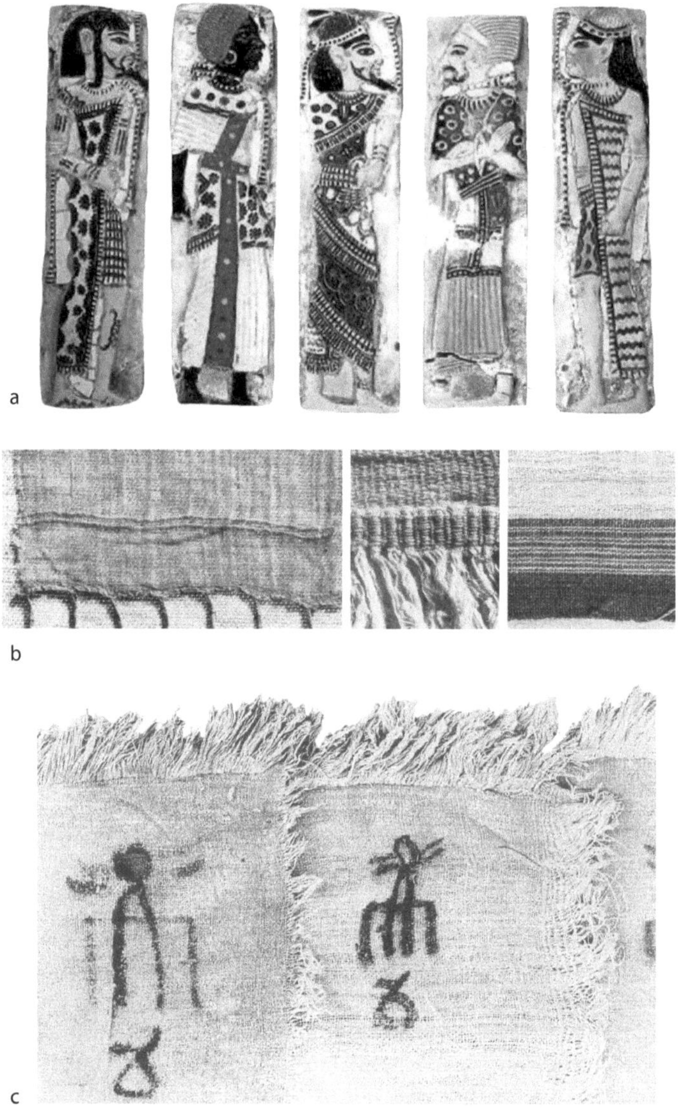

Figure 2.3. Textiles: (a) different dress styles used to mark different ethnic groups on faience plaques from the temple of Ramses III at Medinet Habu (courtesy of Juergen Liepe); (b) edge finishing on Egyptian linen (including, left to right, a self-band, weft fringe and selvedge stripe: van't Hooft et al. 1994: pls. 1.48–9, 2.54, courtesy of the Netherlands National Museum of Antiquities); and (c) inked marks on several Egyptian linen textiles that might refer to the weavers and/or to a local temple (Winlock 1945: pl. xiv. 1–2).

Mediterranean. For example, surviving Egyptian linen garments and cut-cloth reveal distinctive border decoration, fringes, and/or defined selvedges (Figure 2.3b–c; Vogelsang-Eastwood 2000; and note the narrative prominence of these in the Middle Kingdom *Tale of the Eloquent Peasant;* Lichtheim 1975: 169–84).

The use of specialist equipment, fixing agents, dyes, splicing, spinning, and weaving techniques ensured distinctive regional products that knowledgeable consumers could recognise, but other manufacturers would struggle to replicate exactly (see Pinheiro-Machado, this volume; Richardson 2008: 17–18), and Egyptian linen preserves good evidence for some of these technological trademarks (e.g., flax splicing, s-spun thread, certain weaves, and weft fringes: van't Hooft et al. 1994: 13–22; also for the special nature of certain Ugaritan dyes, see van Soldt 1990). There are also preserved examples of Egyptian textile marks, including names, dates, and abstract designs that were woven, embroidered, or written in ink onto the borders of the cloth (Figure 2.3c; see also van't Hooft et al. 1994: Table 6a–c).

A variety of activities, involving both first- and second-hand transactions and tomb robbing, seem to have led to the full range of grades and types of both uncut cloth and finished garments being available for potential purchase (including those of so-called royal quality: Vogelsang-Eastwood 2000: 293; see also Eyre 1998: 178–83; Janssen 1975: 249–98). Mycenaean and Ugaritan documents suggest a similar pattern of distinctive wool and linen products of various grades, colours, and provenances, with decorated fringes and/or selvedges cut to standard lengths or made into recognisable finished garments (e.g., Killen 1979; Palaima 1991: 291–93; Ribichini and Xella 1985; van Soldt 1990).

Finally, the depictions of garments in wall paintings and decorative reliefs typically combine important categorical or ordinal information about the perceived age, gender, ethnicity, or social class of the wearer with great attention to known quality grades and details of weave, colour, edge finishing, and/or diaphony (e.g., Figure 2.3a; Doumas 1992: nos. 100–34). To a degree, however, we should see these categories and ranks as more sharply defined on the wall (and perhaps in death) than they always were in the home or in the street: For example, the velocity with which foreign cloth and foreign garment styles were circulating suggests that, at least for certain occasions, flexible dressing was an important form of social display, particularly for various lower and upper elite sections of society.

Oil and Wine

To a large degree, what the manipulation of textile value reflects is the fact that, unlike metals, textiles could be fashioned almost anywhere in the Mediterranean and Middle Eastern region. What therefore mattered was the ability to monopolise and market particular raw materials, processing techniques, and concentrations of skilled labour. Likewise, the desire to acquire particular types of textile, from particular places, was something to be carefully cultivated, by the development of uniquely recognisable techniques, eye-catching styles and attractive symbolic associations, as well as to be reinforced by protectionist strategies if necessary.

Oil and wine products are also heavily processed commodities that could be produced almost anywhere in the Mediterranean, but the way they were promoted differs from textiles in two crucial ways. First, and to varying degrees, they involved an investment in the landscape that might only have a delayed economic return (e.g., vineyards and olive groves that only produced viable harvests several years or decades after initial planting), but once introduced created reuseable landscape capital (e.g., vines, trees, terraces, processing installations, etc.) that encouraged regional specialisation over the longer term, an accompanying inertia resulting from sunk-costs and, potentially, opportunities for evoking the modern equivalent of terroir. Second, oils and wines were liquids that typically required airtight containers for transport and storage, meaning that: (1) The contents themselves were usually hidden from view until the moment of physical consumption; and (2) The containers became as important a locus for product differentiation as the contents. To some extent, however, this section still awkwardly lumps two very different kinds of commodities and glosses over a host of apparent variation in wine types and additives as well as huge variety of oleaginous plants and additives used for different oils. Even so, the discussion below begins by addressing some areas of common ground between these commodities and then gradually differentiates them.

In the Bronze Age written records, we regularly see oils and wines with specific names, grades, provenances, ingredients, tastes/scents, etc. Both commodities were loosely associated with acts of appropriate hospitality (though in different ways to family, friends, or strangers) and, in many instances, also the health of the living and dead. Although we can certainly point to overland trade of oil and wine over shorter distances, the long-distance bulk transport of such Bronze Age goods only seems to have been worthwhile around the coastal fringes of the eastern Mediterranean and/or down a few main riverine arteries. Oils and wines were, from at least the later 4th millennium BC onwards, transported in containers of increasingly standardised size and shape, but the trade in these commodities increases during the 3rd millennium BC in tandem with the development of sailing ships (with bigger cargo capacities and longer range than paddled craft; see Broodbank 2000: 96–102; Marcus 2002: 409–12) and becomes a standard feature of Mediterranean life through the rest of the Bronze and Iron Age and up to the present day (note, for example, their central role in the modern global branding of a healthy "Mediterranean" lifestyle; see Meneley 2007).

As with textiles, both oils and wines were often used as markers for specific social classes or ethnic groups as they enabled distinctive cooking, eating, drinking, and ablutionary habits. Within these social categories and communities however, they most often forged, on the one hand, either fairly undifferentiated social relationships built on communal smells, age, and gender-related rites, commensal hospitality, or purity-honour-shame idioms (e.g., for the latter in modern olive oil marketing, see Meneley 2007: 683–84), or on the other, a sense of collegiality and reciprocity through cycling obligations of hospitality (e.g., in welcoming people to a banquet: Davies 1973: pls.lxiii–vii) or the equipment deemed necessary to belong to a certain peer group (e.g., oil and wine consumption "sets," see below).

By the mid-2nd millennium BC, the most obvious example of the take-off of seaborne trade is the maritime transport jar that was used for a wide range of commodities, but especially for oil and wine. "Canaanite-style" amphorae are probably the most famous Bronze Age example (Figure 2.4a), but handled amphorae of one type or another were thereafter the key transport containers for the next 2–3,000 years (until they are supplanted by Medieval staved barrels; see Vroom 2003: 15). As Diane Twede has argued (2002, following Lockhart 1997), transport jar design was consistently driven by the three, near-universal demands of commercial packaging: protection of the contents, utility in transport and consumption, and market communication. While the shape of transport jars was usually a physical adaptation to stacking in and extraction from the holds of ships, their approximate weight when full again often hovers around known standards (the talent weight of *ca.* 27–30 kg mentioned above; see, e.g., Heltzer 1990: 127; or half that for the Mycenaean stirrup jars discussed below), which again probably reflects the need for both individuals and donkeys to carry them efficiently upon disembarkation.

Canaanite-style jars were first stoppered, then their necks were covered in clay to make them airtight, and thereafter they were typically sealed and/or marked in some other manner (e.g., Grace 1956: g.2b). The sealing practices are typically known to us only through imagery or accidents of archaeological recovery and marking practices are also very unevenly preserved, with inked, painted, or otherwise biodegradeable labels particularly vulnerable. This seriously skews our impression of the amount of both literate and nonliterate information being conveyed with such containers, especially since the multi-purpose, reusable nature of the jar itself ensured that very few of the more archaeologically robust impressed or incised marks are likely to have referred to a singular contents (for these marks, see Hirschfeld 1993, 2002; unsurprisingly they are far from explicit to the uninitiated). However, it is clear again that the Bronze Age sees the beginning of a practice of marking maritime containers that becomes even more common and more clearly propositional in later periods where it is used to endorse standardised amphorae sizes, contents, producers, provenance, and vintage (e.g., Callender 1965; Eiring and Lund 2004; Grace and Savvatianou-Petropoulakou 1970: 278–80).

The most obviously branded Bronze Age wines and oils are those that we see associated with restricted spheres of royal (and temple) production. In Egypt, for example, while many private vineyards are depicted in tomb paintings or mentioned in documents, the surviving jar labels typically refer to the products from the large estates of the ruler or major temples. What David Wengrow has pointed out for a very early jar label (2008: 9–10, figs. 1–2) is something that we might consider as a fairly persistent form of Egyptian royal product endorsement: the ruler as both action hero (defeating enemies and running from one corner of the kingdom to another) and fertility figure (giving life to the land; for a modern example with similar tension between roles; see also Holt and Thompson 2004).

Such royal jar labels (and less elaborate ones) certainly imply some indirect benefit to be gained by those who were lucky enough to share in the royal production, but did they ever circulate beyond the seemingly closed loop of royal gifts? In fact, there is a range of evidence to suggest that they did, at least by the later 2nd millennium BC if not before, and in both first- and second-hand commercial transactions (see below for Ugarit and the Aegean). For Egypt, the situation may initially seem unclear, but we can certainly document the flow of such commodities on the second-hand (and in at least some cases unopened) market, often after first-hand disbursement by the palace and/or temples (Tallet 1998: 260–61). For example, the workers at Deir el-Medina during the 13th–11th centuries BC seem to have occasionally been given *neheh* oil (almost certainly olive oil; see Tallet 2004) by royal allocation on particular festivals (unsurprising given the special nature of this community working on royal tomb projects), but some jars are then also exchanged as commodities by the workers amongst themselves (Janssen 1975: 330–42, 350–52).

The case of *neheh* oil also gives us a rare opportunity to consider a full product trajectory, from manufacture to consumption. The oil was clearly used for a range of purposes and 14th–12th century BC documentary evidence suggests that at that time, a litre was typically worth about 200 g of copper or 3 g of silver, whereas wine was both rarer and more costly (following Janssen suggested figures, 1975: 108–09, 330 n.6). However, the dating of the jar labels and other documentary references indicate some important shifts in likely provenance: The earlier examples within the above date range suggest strongly that the contents were imported and microscopic fabric analysis of the actual jars suggests that these were made near the Lebanese and Syrian coasts (Serpico 2004; Smith et al. 2004).

Within this region, the large, cosmopolitan port town of Ugarit is likely to have been one of the major centres of production, especially in the light of the vivid evidence this site has provided for advanced methods of olive oil manufacture (e.g., the lever beam press: Callot 1987), arrangements for bulk export (e.g., harbourside stores of large groups of jars; see Figure 2.4a), the celebration of oils and wines in annual religious events (e.g., Lipinski 1988: 140–42) and their obvious commercial interest to palace, temples, and private individuals alike (Bevan 2007: 145–50; Heltzer 1987; Nougayrol et al. 1968: 80–83; Schaeffer 1949: pl.31). However, by the latter half of the 13th century BC, this foreign oil is increasingly replaced by local products from the western Delta: Not only are the jars now often made of local Egyptian clays, but the labels frequently now claim that the oil is "made in Egypt," and one contemporary document notes the coexistence of both Syrian and Egyptian varieties (Tallet 2004: 64–67).

This kind of unusually detailed and converging evidence suggests clear patterns of both production-for-export and deliberate import substitution, just as it also indicates combinations of structured institutional disbursement and commercial circulation. More generally, Bronze Age oils and thicker unguents were made from a range of different oleaginous plants, and different types and grades are an obvious feature amongst the documentary records (Heltzer 1987; Leemans 1960: 14–16; Melena 1983: 109–17; Serpico and White 2000: 390–406; Shelmerdine 1985: 17–39). Oils were used as sources of light, smoke, and general atmosphere in living spaces and they were also comestibles; but the most obvious Bronze Age uses were as perfumed bodily treatments (and related to this, sometimes also as treatments of textiles to make them shine and smell like the bodies of the gods; see Shelmerdine 1995). The finest perfumed oils were incredibly complex recipes, made of a mixture of ingredients, whose proper manufacture took many days and required the skills of expert perfumers (e.g., Shimy 1997: 315–67). Perhaps the best example of the elaborate consumption choreography that might be associated with these products are Egyptian prescriptions for "seven sacred oil" sets that were both for anointing living bodies and to allow the deceased to pass through each of the seven gates of the underworld (Gee 1998: Table 7.5; Shimy 1997: 26–119). These oils could be made of purely Egyptian ingredients, purely foreign ones, or a combination, and by the later 3rd millennium at least, each had been given a specific name, place in the anointing ceremony and physical slot in the oil sets themselves, providing a very clear hierarchical ordering of space and time, but ultimately, also a ranking of social relationships amongst the living and the dead.

Beyond this example of upper elite, hyper-crafted oils (and their lower elite simulacra: Roth 1992: fig. 3), there were also a range of others that still involved complex preparation, but were produced in greater quantities

Figure 2.4. Oils, wines, and related liquid commodities: (a) a deposit of over 80 Canaanite-style transport jars from the northern harbour of Ugarit (Schaeffer 1949: pl. ix); (b) a large coarseware stirrup jar painted with Linear B signs (Demakopoulos 1981: 22, courtesy of the 9th Ephorate of Prehistoric and Classical Antiquities, Thebes); (c) a small stirrup jar; (d) a base-ring juglet; and (e) a mixing bowl decorated with a chariot scene. All images not otherwise attributed are courtesy of the British Museum.

and circulated throughout much of the eastern Mediterranean. Broadly speaking, we can distinguish two scales to the trade in oils that first appears in the 3rd millennium, but becomes very clear-cut by the middle of the 2nd millennium BC, one carried out in large coarseware containers and another in decorated juglets. Our degree of clarity about the relationship between these two scales varies with region and context, but the overall suspicion is that the smaller vessels contained more highly processed oils (and other precious liquid products not all of which need have been oil based), allowing particularly well-connected trading centres to add value to broader regional products through elaborate repackaging, extra physical ingredients and perhaps various kinds of ritual endorsement.

While the discussion below returns to one particularly interesting example of product labelling on coarseware containers, it is the smaller, more decorated versions that show the strongest signs of having been heavily marketed commodities. From the middle of the 2nd millennium BC onwards, we see a truly impressive range of decorated juglets being made throughout much of the eastern Mediterranean (typically less than half a litre in volume, but with lots of diversity by type).[4] These objects exhibit very high levels of stylistic innovation, a range of cultural references (regional styles, different surface treatments and technical allusions), and a tendency towards being substituted for local imitations (e.g., Killebrew 2004; Sherratt 1999).

Their different shapes choreographed pouring in diverse ways and are likely to have had both a gestural and a linguistic vocabulary of appropriate delivery (e.g., the grip taken on differently placed handles, the size and positioning of the spout, with a differentiated terminology similar to, for example, pour, glug, drizzle, or strain in modern English). Moreover, although there was huge variety in juglet styles, most styles exhibit reasonable internal consistency, suggesting that they were markers for equally recognisable, standardised contents (that need not, of course, all have been oils). Specific styles also seem to have positioned their contents for particular regional markets or particular kinds of consumer: Some may have evoked the contents directly, the most famous but contentious example being those closely resembling a slashed poppy-head because their contents may have been opium based ("base-ring" juglets; see Figure 2.4d; Bisset et al. 1996; Merrillees 1962). Others were metal skeuomorphs and thus intimately connected with class-related aspirations to use higher value metal vessels (Bevan 2007: 136–37), or were burnished-and-incised and thus closer to self-consciously traditional organic designs (e.g., from wood, gourds, basketry, or leather) or perhaps even practices of personal body-marking (e.g., tattoos or body-paint).

The Mycenaean "stirrup-jar" is a good example of the juglet phenomenon (Figure 2.4c). Over perhaps a 400-year period in the later Bronze Age, its external painted decoration is repeatedly modified, with both simple linear designs and sometimes figurative imagery. Some of this variation is regional and contemporary in character, but much of it is chronological, reflecting a high stylistic turnover in decorated Mycenaean pottery more generally that, in modern archaeological terms, has made it one of the finest-grained archaeological indicators for relative chronology in all of world prehistory. The single, narrow spout on stirrup-jars (without a second airhole) makes oil drip out very slowly, and the two strap handles wrapped over a false spout are visually unusual and differentiate the physical act of pouring from that of ordinary juglets. Stirrup jars were also made in much larger, less decorated coarseware versions (Figure 2.4b, with a typical capacity of 12–14 litres), and these seem to have been the main liquid transport jar in the late 2nd-millennium Aegean (with oil assumed to be one of the most common contents). The large coarseware versions were a scaling-up of the small, fineware juglets and were arguably less well-adapted to being a maritime transport shape than, for example, the Canaanite-style jar (the latter certainly traded to the Aegean, but was not locally made there and was probably full on arrival rather than traded for its own sake), but they capitalised on the familiar and distinctive qualities of the smaller version.

Although many of the small, decorated stirrup-jars that travelled to the eastern Mediterranean were made in the Argolid, the clays of the larger coarseware transport jars suggest that western Crete was a major production region (Haskell 2004). Large numbers are mentioned in the Linear B documents (e.g., in at least one transaction of 1,800 jars; see Shelmerdine and Bennet 2008: 304), and a group was found stoppered and sealed in a (possible merchant's) store at Mycenae, seemingly ready for onward consignment (Tournavitou 1995: 79–81, pls. 11–12a). Such large coarseware versions certainly were traded beyond the Aegean to both the eastern and central Mediterranean, but are less easily identified in excavations than the more vividly decorated juglets, hence it remains difficult to assess how much breaking-of-bulk was done within the Aegean first. The probable role of Cypriot distributors as well as locals seems likely, however, as many of the jars show incised marks on their handles that overlap with those of Cypriot imitations and seem similar in concept to the marks on copper oxhide ingots (Hirschfeld 1993, 2002).

Over 180 such coarseware transport jars also have large painted Linear B signs across the body or on the shoulder (Figure 2.4b; Catling et al. 1980; Haskell 2004), referring to the manufacturer (of the pot and/or of the contents), a probable distributor or owner of some kind, and/or a place name (of manufacture or distribution). Both the clays and the place names suggest that these jars were made in both south-central and western Crete, while the archaeological findspots of these vessels suggest strikingly directional patterns of exportation with each of these two regions supplying different mainland centres (Palaima 1984: 191–94). It is tempting to suggest that bulk oil from Crete was being shipped to the mainland, where it was then enhanced and repackaged for onwards trade in the smaller decorated stirrup jars, but it is worth noting that western Crete was also producing extremely distinctive finewares for export at this time (stirrup-jars for smaller amounts of oil, rhyta for straining liquids, deep bowls for mixing wine; see Kanta 1980: 288–89; Tzedakis 1969), so a more complex picture of the combined marketing of both more and less refined products by several different regional centres seems more likely.

On several of the labelled coarseware jars, the name of the collector/owner is replaced by the adjective "royal." And, as with the textiles and other oils discussed above, we are left to decide whether this implies actual royal control and ownership of the product or something more propositional about the perceived quality of the contents. The abbreviated formulae used for these labels certainly share much in common with those found in the Linear B palace records that deal with bulk commodity transactions (for an overview, van Alfen 1997). However, a traditional interpretation of palatial redistribution, although indicating a key feature of commodity flows, nonetheless frames our understanding of palatial and nonpalatial action in an entirely unsatisfactory way, for at least five reasons.

First, the archives are incredibly partial, reflecting only the subjects recorded in clay, and of those, only the ones preserved by accidental firing. They also reflect only the palace's short-term interest in a very limited range of raw materials, semi-finished, and finished goods (e.g., metals and metalwork, processed oils, textiles, and the ingredients necessary to make them). In fact, there is very little coverage of whole swathes of a Mycenaean kingdom's political, literary, legal, diplomatic, or economic life (e.g., Palaima 1991; and for the importance of assessing these kind of issues of preservation and administrative reach in other Mediterranean and Middle Eastern archives, see Postgate 2001). In fact, the abbreviated vocabulary of palatial administration may well have permeated nonpalatial commodity transactions (as we can see that they did elsewhere in the eastern Mediterranean) but were just less often recorded on clay. Second, the labelled stirrup jars themselves may date to a period after the final collapse of the Mycenaean palace at Knossos and, if true, must be decoupled partially or wholly from models of (at least pan-Cretan) royal administration. Third, oil seems to have been a very secondary concern (to wool and textile production) for the Knossos palace in any case, and it would have been able to acquire the amounts mentioned in the archives from its immediate hinterland (Palaima 1984: 201–03), indicating the archaeologically documented scale of oil trade cannot simply be mapped onto palatial production. Fourth, the large size and prominent position of the labelling in a place that was best seen when the jars were not in a hold or stored en masse (i.e., in contrast to the incised marks on the tops of the handles) suggests that their purpose was partly or wholly as a visual enticement. The fact that, in one or two instances, the Linear B signs are so poorly done that the painter may not have been fully aware of their meaning argues in the same direction. Fifth and finally, the Knossos Linear B archive shows the palace acquiring resources from more distant parts of Crete, indirectly, through people that have been given a modern gloss as "collectors/owners" (Bennet 1992). Although the documents certainly do not provide clear-cut evidence about the identity of such people, nevertheless some, if not all, were from the upper elite class, were semi-independent in their operations, and were at least potentially similar to elite traders documented elsewhere in the eastern Mediterranean (Killen 1979: 176–79). What this extended example should also be particularly reminiscent of is the trade in copper, where there are all sorts of coals-to-Newcastle conditions, efforts to endorse qualities and quantities by a range of means, markings at various stages of the production and distribution process, co-marketing of both hyper-elaborate finished goods and less processed versions, possible Cypriot involvement, and mixed political, economic, and religious associations.

The examples above have focused on cases where oil is the most likely contents, but it is worth briefly returning to examples from the wine trade. One general difference between modern wine and that consumed in the Bronze Age is that mixing and removing various additives was a far more important part of Bronze Age wine production, distribution, and consumption routines. Another possible difference is the modern emphasis of curating and ageing wines, with all of the opportunities for biographies of past ownership that this offers (Silverstein 2006: 484). The Bronze Age evidence for the latter practice is equivocal (but see Meeks 1993: 25–26 for a possible 35-year-old wine from the tomb of Thutankhamun, and for a later Homeric tradition of aged and preowned wine, see *Iliad* VII.467–75, Odyssey II.340, III.391–92, IX.196–215), but they were clearly stored over at least several years and offered the same kinds of opportunities for second-hand circulation (especially for "royal" products), as the *neheh* oil discussed above.

In Egypt, wine had been imported since perhaps the late 4th millennium, particularly from the area of modern-day Israel-Palestine; but by the middle of the 2nd millennium BC, production within the Nile delta expanded dramatically (Bourriau 2004; McGovern 2003: 85–147). Evidence for both substantial royal and private vineyards exists, but large quantities of wine were also rebottled and relabelled for royal jubilees and other festivals (e.g., McGovern 1997). In a sense, this is a classic example of royal redistribution, but we risk cutting out an important part of these objects' attraction if we ignore their postdisbursement afterlife as royally endorsed products. Egypt's good conditions for archaeological preservation also mean that here, unusually, we have evidence for Egyptian-made, but Canaanite-style transport jars, marked and sealed in a variety of ways including pre-firing stamps, ink labels, and mud cones to make the vessels airtight. The legible marks suggest that wine was by far the most common contents in such jars (though see also the *neheh* oil discussed above) and also often indicate the source and destination for the products, its vintage, vintner, and quantity (e.g., Cerny 1965; Hayes 1951). Different varieties of wine and different quality grades are clearly visible in contemporary New Kingdom Egypt, and such intensification of consumption and diversification of taste also brought with it a variety of novel consumer habits such as decanting into a shallow bowl or sipping through a straw (Meeks 1993: 26–27).

Local eastern Mediterranean wine industries rose and fell as they competed for popularity along the strip from the Egyptian delta to Syria. Given equal opportunities for production in these regions (though less true further east in lowland Mesopotamia where conditions were less favourable; see Powell 1995), it was the ability of producers to cultivate particularly attractive products and a sense of tradition that was important. It is no accident that we see elaborate wine festivals at Ugarit, whose wine production seems to have dwarfed even its production of oil (Lipinski 1988: 140–42). These rituals involved the distribution of large quantities of wine, the involvement of both royal and divine patrons, and the deliberate promotion of Bronze Age terroir, with one poetic text, for example, evoking a "necklace" of terraced vineyards that ringed the Lebanese mountains and were endorsed by the father of the gods, El himself (Watson 1999; for the suggestion that these vineyards were terraced, see Heltzer 1990: 120).

By contrast, although Cretan and Mycenaean palatial states produced a range of wines and traded these within the Aegean (including single transactions of as much as 12–14,000 litres, as well as a range of logo-like, ligatured ideograms, and other adjectival product distinctions; see Palmer 1994, 2000), there is not much evidence for these products circulating more widely in the eastern Mediterranean until the Iron Age. However, what certainly were very popular in the eastern Mediterranean during 14th–12th centuries BC were highly decorated Mycenaean wine-mixing bowls ("pictorial kraters"; see Figure 2.4e). By this time, metal wine sets were must-have paraphernalia for elite families as a means of confirming their membership of a particular social rank, and within this, of promoting reciprocal behaviour amongst their perceived peers. Pottery versions were, by contrast, some way further down the value hierarchy, and although sometimes also found in high-status contexts, were not formally worthy of mention in royal circles for example. As Sue Sherratt puts it nicely (1999: 195), such pottery therefore "has both less importance and more importance than has often been accorded it": less because it is not something in which the palaces and temples ever expressed any really interest (Whitelaw 2001; in contrast to their high modern valuation by art historians, archaeologists, or the antiquities market) and more because: (1) Both producers and traders clearly valued them enough to trouble to make, copy and distribute them; and (2) The extensive distribution of these vessels restates the fundamental point that transregional trade included a substantial lower and sub-elite component.

In any case, these mixing bowls were probably first made in the north-western Peloponnese of Greece (Schallin 1997), but are in fact more commonly found in export contexts elsewhere in the eastern Mediterranean and seem to have been produced with an export market in mind. Indeed, their elaborate decorative scenes of chariot riding, bulls, and wildfowl deliberately evoked a common elite lifestyle throughout much of the north-eastern Mediterranean. There are many other instances of such marketing of apparently nonpalatial pottery products for external markets (the juglets mentioned above being another excellent case), reflecting important feedback between producers, distributors, and consumers. A further sign of how competitive such a market could be is the fact that by the 13th–12th centuries BC, near identical mixing bowls were being imitated on Cyprus as local substitutes (Sherratt 1982), to the extent that both import and imitation sometimes appear alongside each other, and local imitations eventually supplant them entirely.

Conclusion

The above discussion has inevitably mixed general summary with selective detail and no doubt missed many important case studies, but it has sought to explore the making and marking of four Bronze Age commodities—metals, textiles, oils, and wine—at an early stage in their development as familiar components of Mediterranean life. In the Bronze Age, substitutable goods evoked a whole range of quality and quantity distinctions and circulated in networks of gifting, commercial exchange, and second- or third-hand transaction that are not always easy to disentangle. However, we do little to further our understanding of the values assigned to such objects, in Bronze Age society or any other, either by uncritically equating them with modern branded goods or by modelling their world as a premodern, non-Western other. In fact, just as capitalism as an analytical concept is better divorced from previously wedded notions such as Western democracy or the Protestant work ethic, so, too (and as this edited volume suggests more generally), the concept of commodity branding is better decoupled from any automatic associations with postindustrial, Western capitalism.

To conclude, I would like to suggest two broad areas in which those different regional, thematic, or historical research programmes seeking to address commodities can work most effectively. The first is cross-cultural attention to different scales of brand and different branding agents. Brands themselves exist at various levels of abstraction, from loose ideas and lifestyles, to specific names and places, to the marks on physical products. They can be promulgated by producers, distributors, or consumers and this may involve any combination of individuals, families, factions, communities, corporations, nation-states, and/or divine figures. Therefore, who proposes the new social contexts into which substitutable goods might fit and how is this agency conceived in ideological and practical terms? For modern brands, early analytical emphasis was understandably placed on the defining role of the corporation and of certain kinds of media specialist who were cultural intermediaries (e.g., Negus 2002), but the existence of cheap, fast, global communication networks, and widespread literacy in much of the modern world arguably opens up a whole range of other possibilities that we are only just beginning to comprehend.

In the Bronze Age, the only entities that seem to have had the capacity regularly to span long-range and otherwise attenuated chains of economic interaction were rulers, certain temple institutions, and the upper elite administrators and traders, and even they did so tenuously and with a bravado that was not always matched by either their practical knowledge of the links in the chain or the forces of commodification and decommodification that they were unleashing. Even so, gods and rulers were key authenticators of Bronze Age value, whether or not the goods involved moved in immediately commercial circuits. Indeed, as with all branded commodities, it was the existence of singular opposites—the goods that the palaces and temples temporarily kept back from circulation—that was arguably crucial (for an excellent ethnographic example of how royal intervention of this kind might be structured, see Warnier, this volume).

A second area of potential common ground is the one raised at the beginning of this chapter: the need for greater attention to the culturally specific but structurally equivalent models through which people coordinate their social relationships. Relational models frequently provide the operational framework by which goods are commodified in the first place, and by which they are later given fresh social meaning. In the Bronze Age Mediterranean and Middle East, social and economic exchanges of all kinds worked within the conceptual framework of appropriate conduct in families and small villages to the extent that even the most seemingly generic commodities were made and marked in ways that cued for their later reintegration into society in a set number of ways. Indeed, without such cues, Bronze Age trade, with its unfamiliar divine and royal endorsements, seemingly irrational behaviours, and endless owls-to-Athens, would make little sense.

Notes

1. Acknowledgements: My particular thanks to David Wengrow and the participants of the Anthropological and Archaeological Approaches to Commodity Branding conference for their input. In addition, John Bennet, Cécile Michel, Oliver Pryce, Sue Sherratt, and John Tait offered very useful advice on particular topics or commented on draft versions. My thanks also to the anonymous reviewers of this volume for their helpful suggestions.
2. A wider comparison between the Cypriot and Dilmun/Magan metal trading regimes is the subject of graduate research by Helen Crossman (2007 and an ongoing doctoral project at the University of Reading).
3. Indeed, in more modern cases where particularly prestigious cloth is used, the selvedge is often prominently displayed in the final garment and/or made into a hem (G. Clark, this volume).
4. This juglet phenomenon is only mentioned briefly here but was considered in greater detail by Lesley Bushnell at the conference from which this volume emerged. It is the subject of her doctoral research (University College London) and will be published elsewhere.

References

Aaker, J. L. 1997. Dimensions of brand personality. *Journal of Marketing Research* 34:347–56.

Aggarwal, P. 2004. The effects of brand relationship norms on consumer attitudes and behavior. *Journal of Consumer Research* 31:87–101.

Akerlof, G. 1970. The market for "lemons": Quality uncertainty and the market mechanism. *The Quarterly Journal of Economics* 84:488–500.

Alberti, M. E. and N. Parise. 2005. Towards a unification of mass-units between the Aegean and the Levant. In R. Laffineur and E. Greco (Eds.), *Emporia. Aegeans in the Central and Eastern Mediterranean*: 382–91. Liege, Belgium: University of Liege.

Appadurai, A. 1986. Introduction: Commodities and the politics of value. In A. Appadurai (Ed.), *The Social Life of Things*: 3–63. Cambridge: Cambridge University Press.

Arvidsson, A. 2005. Brands: A critical perspective. *Journal of Consumer Culture* 5:235–58.

Aswani, S. and P. Sheppard. 2003. The archaeology and ethnohistory of exchange in precolonial and colonial Roviana. *Current Anthropology* 44 Supplement:51–78.

Aufrère, S. 1991. *L'Univers Minérale dans la Pensée Egyptienne*. Paris: Institut Français d'Archéologie Orientale du Caire.

Bass, G. F. 1967. Cape Gelidonya: A Bronze Age shipwreck. *Transactions of the American Philosophical Society* 57:1–177.

Bennet, J. 1992. "Collectors" or "owners"? An examination of their possible functions within the palatial economy of LMIII Crete. In J.-P. Olivier (Ed.), *Mykenaïka. Actes du IXe Colloque International sur les Textes Mycéniens et Egées Organisé pa le Centre de l'Antiquité Greque et Romaine de la Fondation Héllenique des Recherches Scientifiques et l'Ecole Française d'Athenes*: 65–101. Paris: Boccard.

Bennet, J. 2008. Palace™: Speculations on palatial production in Mycenaean Greece with (some) reference to glass. In C. Jackson and E. E. Wager (Eds.), *Vitreous Material in the Late Bronze Age Aegean: A Window to the East Mediterranean World*: 151–72. Oxford: Oxbow (Sheffield Studies in Aegean Archaeology 8).

Bevan, A. 2007. *Stone Vessels and Values in the Bronze Age Mediterranean*. Cambridge: Cambridge University Press.

Beverland, M. B. 2005. Crafting brand authenticity: The case of luxury wines. *Journal of Management Studies* 42:1003–29.

Bisset, N. G., J. G. Bruhn, S. Curto, B. Holmstedt, U. Nyman, and M. H. Zenk 1996. Was opium known in 18th Dynasty Egypt? An examination of materials from the tomb of the Chief Royal Architect Kha. *Agypten und Levante* 6:199–201.

Blackman, A. M. and T. E. Peet. 1925. Papyrus Lansing: A translation with notes. *Journal of Egyptian Archaeology* 11:284–98.

Bourriau, J. 2004. The beginnings of amphora production in Egypt. In J. Bourriau and J. Phillips (Eds.), *Invention and Innovation. The Social Context of Technological Change 2. Egypt, the Aegean and the Near East 1650–1150 BC*: 78–95. Oxford: Oxbow.

Broodbank, C. 2000. *An island archaeology of the Early Cyclades*. Cambridge: Cambridge University Press.

Callender, M. H. 1965. *Roman Amphorae*. Oxford: Oxford University Press.

Callot, O. 1987. Les Huileries du Bronze Recent à Ougarit. In M. Yon (Ed.), *Le Centre de la Ville. Ras-Shamra-Ougarit III. 38e–44e Campagnes (1978–1984)*: 197–212. Paris: Etudes Recherche sur les Civilisations.

Castle, E. W. 1992. Shipping and trade in Ramesside Egypt. *Journal of the Economic and Social History of the Orient* 35:239–77.

Catling, H. W., J. F. Cherry, R. E. Jones, and J. Killen. 1980. The Linear B inscribed stirrup jars and West Crete. *Annual of the British School at Athens* 75:49–113.

Cerny, J. 1965. *Hieratic Inscriptions from the Tomb of Tutankhamun*. Oxford: Oxford University Press.

Charpin, D. and J.-M. Durand. 1991. La suzeraineté de l'empereur (Sukkalmah) d'Elam sur al Mésopotamie et le "nationalisme" Amorrite. In L. De Meyer and H. Gasche (Eds.), *Mésopotamie et Elam*: 61–66. Ghent, Belgium: University of Ghent.

Cierny, J. and G. Weisgerber. 2003. Bronze Age tin mines in Central Asia. In A. Giumlia-Mair and F. LoSchiavo (Eds.), *The Problem of Early Tin*: 23–34. Oxford: Archaeopress.

Crossman, H. 2007. A comparative study of Cyprus and Bahrain, and the Bronze Age trade of copper in the Eastern Mediterranean and Persian Gulf. Master's thesis, University College London.

Cutler, B. and J. Macdonald. 1977. The unique Ugaritic Text UT 113 and the question of "guilds." *Ugarit-Forschungen* 9:13–30.

Davies, N. De G. 1935. The work of the graphic branch of the expedition. *Bulletin of the Metropolitan Museum of Art* 30:49–52.

Davies, N. De G. 1973. *The Tomb of Rekh-mi-re at Thebes*. New York: Arno Press.

de Maaijer, R. 2001. Late Third Millennium identifying marks. In W. H. van Soldt (Ed.), *Studies Presented to Klaas R. Veenhof on the Occasion of His Sixty-Fifth Birthday*: 300–24. Leiden: Netherlands Institute for the Near East.

Demakopoulou, K. 1981. *The Archaeological Museum of Thebes Guide*. Athens: General Director of Antiquities.

Dercksen, J. G. 1996. *The Old Assyrian Copper Trade in Anatolia*. Istanbul: Nederlands Historisch-Archaeologisch Instituut.

Doumas, C. 1992. *The Wall-Paintings of Thera*. London: Thera Foundation.

Durand, J.-M. 1988. *Archives Epistolaires de Mari, Volume I.1 of Archives de Mari 26.1*. Paris: Editions Recherche sur les Civilisations.

Eiring, J. and J. Lund, Eds. 2004. *Transport Amphorae and Trade in the Eastern Mediterranean. Acts of the International Colloquium at the Danish Institute at Athens*. Athens: The Danish Institute at Athens.

Eyre, C. J. 1998. The market women of Pharaonic Egypt. In N. Grimal and B. Menu (Eds.), *Le Commerce dans l'Egypte Ancienne*: 173–91. Paris: Institut Français d'Archéologie Orientale.

Fanselow, F. S. 1990. The bazaar economy or how bizarre is the bazaar really. *Man* 25:250–65.

Feuerbach, A. 2006. Crucible Damascus steel: A fascination for almost 2,000 years. *Journal of the Minerals, Metals and Materials Society* 58:48–50.

Fiske, A. P. 1991. *Structures of Social Life: The Four Elementary Forms of Human Relations*. New York: The Free Press.

Fiske, A. P. 2000. Complementarity theory: Why human social capacities evolved to require cultural complements. *Personality and Social Psychology Review* 4:76–94.

Fiske, A. P. 2004a. Four modes of constituting relationships: Consubstantial assimilation; space magnitude, time and force; concrete procedures; abstract symbolism. In N. Haslam (Ed.), *Relational Models Theory: A Contemporary Overview*: 61–146. Mahwah, NJ: Erlbaum.

Fiske, A.P. 2004b. Relational models theory 2.0. In N. Haslam (Ed.), *Relational Models Theory: A Contemporary Overview*: 3–25. Mahwah, NJ: Erlbaum.

Fiske, A. P. and N. Haslam. 2000. The four basic social bonds: Structures for coordinating interaction. In M. Baldwin (Ed.), *Interpersonal Cognition*: 267–98. New York: Guilford.

Foster, B. R. 1977. Commercial activity in Sargonic Mesopotamia. *Iraq* 39:31–43.

Foster, R. J. 2005. Commodity futures. Labour, love and value. *Anthropology Today* 21:8–12.

Foxvog, D. A. 1995. Sumerian brands and branding irons. *Zeitschrift für Assyriologie und Vordasiatische Archäologie* 85:1–7.

Fuller, T. 1840. *A History of the Worthies of England*, Volume 2. London: Thomas Tegg.

Gale, N. H. 1991. Copper oxhide ingots: Their origin and their place in the Bronze Age metals trade in the Mediterranean. In N. H. Gale (Ed.), *Bronze Age Trade in the Mediterranean*: 197–240. Jonsered, Sweden: Paul Aström.

Gee, J. L. 1998. The requirements of ritual purity in ancient Egypt (PhD dissertation, Yale University).

Goody, J. 2006. *The Theft of History*. Cambridge: Cambridge University Press.

Gordon, C. H. 1956. Ugaritic guilds and Homeric Demiorgoi. In S. S. Weinberg (Ed.), *The Aegean and the Near East: Studies Presented to Hetty Goldman on the Occasion of Her Seventy-Fifth Birthday*: 136–43. Locust Valley, NY: Augustin.

Goren, Y., S. Bunimovitz, I. Finkelstein, and N. Na'aman. 2003. The location of Alashiya: New evidence from petrographic investigation of Alashiyan tablets from El-Amarna and Ugarit. *American Journal of Archaeology* 107:233–55.

Grace, V. R. 1956. The Canaanite jar. In S. S. Weinberg (Ed.), *The Aegean and the Near East: Studies Presented to Hetty Goldman on the Occasion of Her Seventy-Fifth Birthday*: 80–109. Locust Valley, NY: Augustin.

Grace, V.R. and M. Savvatianou-Petropoulakou. 1970. Les timbres amphoriques Grecs. In P. Bruneau, C. Vatin, U. Bezerra, and D. Meneses (Eds.), *Exploration Archéologique de Délos Faite par l'Ecole Française d'Athènes. L'Ilot de la Maison des Comédiens*, Volume 25: 277–382. Paris: Boccard.

Graziadio, G. 2003. Il lingotti ox-hide nella glittica cipriota. *Studi Micenei ed Egeo-Anatolici* 45:27–69.

Hamilton, G. G. and C.-K. Lai. 1989. Consumerism without capitalism: Consumption and brand names in late imperial China. In H. J. Rutz and B. S. Orlove (Eds.), *The Social Economy of Consumption*: 253–79. New York: University Press of America.

Haskell, H. W. 2004. Regional trade patterns in Mycneaean Crete. In S. A. Immerwahr and A. P. Chapin (Eds.), *Charis. Essays in Honor of Sara A. Immerwahr*: 151–60. Athens: American School of Classical Studies at Athens.

Haslam, N. 2004. Research on relational models: An overview. In N. Haslam (Ed.), *Relational Models Theory: A Contemporary Overview*: 61–146. Mahwah, NJ: Erlbaum.

Hauptmann, A., R. Maddin, and M. Prange. 2002. On the structure and composition of copper and tin ingots excavated from the shipwreck of *Uluburun*. *Bulletin of the American School of Oriental Research* 382:1–30.

Hayes, W. C. 1951. Inscriptions from the palace of Amenhotep III. *Journal of Near Eastern Studies* 10:35–56.

Heath, D. and A. Meneley. 2007. Techne, technoscience, and the circulation of comestible commodities: An introduction. *American Anthropologist* 109:593–602.

Hellbing, L. 1979. *Alasia Problems.* Gothenburg, Sweden: Paul Aström.

Heltzer, M. 1982. *The Internal Organization of the Kingdom of Ugarit.* Wiesbaden, Germany: Ludwig Reichert.

Heltzer, M. 1984. Private property at Ugarit. In A. Archi (Ed.), *Circulation of Goods in Non-Palatial Context in the Ancient Near East*: 161–94. Rome: Edizioni dell'Atene.

Heltzer, M. 1987. Olive growing and olive oil in Ugarit. In M. Heltzer and D. Eitam (Eds.), *Olive Oil in Antiquity, Israel and Neighbouring Countries from Neolith to Early Arab Period*: 106–20. Padua, Italy: Sargon.

Heltzer, M. 1990. Vineyards and wine in Ugarit. *Ugarit-Forschungen* 22:119–35.

Hirschfeld, N. 1993. Incised marks (post-firing) on Aegean wares. In C. Zerner (Ed.), *Wace and Blegen. Pottery as Evidence for Trade in the Aegean Bronze Age*: 309–18. Amsterdam: J. C. Gieben.

Hirschfeld, N. 2002. Marks on pots. Patterns of use in the archaeological record at Enkomi. In J. S. Smith (Ed.), *Script and Seal Use on Cyprus in the Bronze and Iron Ages*: 49–109. Boston: Archaeological Institute of America.

Holt, D. B. 2002. Why do brands cause trouble? A dialectical theory of consumer culture and branding. *Journal of Consumer Research* 29:70–90.

Holt, D. B. and C. J. Thompson. 2004. Man-of-action heroes: The pursuit of heroic masculinity in everyday consumption. *Journal of Consumer Research* 31:425–40.

Janson, A. 2002. The mediatization of consumer culture.Towards an analytical framework of image culture. *Journal of Consumer Culture* 2:5–31.

Janssen, J. J. 1975. *Commodity Prices from the Ramesside Period: An Economic Study of the Village of the Necropolis Workers.* Leiden, Netherlands: Brill.

Joannes, F. 1991. L'Etain de l'Elam à Mari. In L. De Meyer and H. Gasche (Eds.), *Mésopotamie et Elam*: 65–76. Ghent, Belgium: University of Ghent.

Kanta, A. 1980. *The Late Minoan III Period in Crete. A Survey of Sites, Pottery and Their Distribution.* Gothenburg, Sweden: Paul Aström (Studies in Mediterranean Archaeology 58).

Karageorghis, V. and G. Papasavvas. 2003. A bronze ingot-bearer from Cyprus. *Oxford Journal of Archaeology* 20:339–54.

Kassianidou, V. and A. B. Knapp. 2005. Archaeometallurgy in the Mediterranean: The social context of mining and trade. In E. Blake and A. B. Knapp (Eds.), *The Archaeology of Mediterranean Prehistory*: 252–69. Oxford: Blackwell.

Killebrew, A. E. 2004. New Kingdom Egyptian-style pottery in Canaan: Implications for Egyptian rule in Canaan during the 19th and early 20th Dynasties. In G. N. Knoppers and A. Hirsch (Eds.), *Egypt, Israel and the Ancient Mediterranean World*: 309–43. Leiden, Netherlands: Brill.

Killen, J. 1979. The Knossos Ld(1) Tablets. In E. Risch and H. Mühlestein (Eds.), *Colloquium Mycenaeum*: 151–81. Geneva: Librairie Droz.

Knapp, A. B. 1986. *Copper Production and Divine Protection: Archaeology, Ideology and Social Complexity on Bronze Age Cyprus*. Gothenburg, Sweden: Paul Aström.

Komter, A. 2001. Heirlooms, nikes and bribes: Towards a sociology of things. *Sociology* 35:59–75.

Kopytoff, I. 1986. The cultural biography of things: Commoditisation as a process. In A. Appadurai (Ed.), *The Social Life of Things: Commodities in Cultural Perspective*: 64–91. Cambridge: Cambridge University Press.

Kramer, S. N. 1963. *The Sumerians: Their History, Culture and Character*. Chicago: Phoenix.

Lagarce, J., E. Lagarce, A. Bounni, and N. Saliby. 1983. Les Fouilles a Ras Ibn Hani en Syrie (Campagnes de 1980, 1981, et 1982). Contribution à L'Etude de Quelques Aspects de la Civilisation Ugaritique. *Comptes Rendus de l'Academie des Inscriptions and Belles-Lettres* 74: 249–90.

Larsen, M. T. 1977. Seal use in the Old Assyrian period. In M. Gibson and R. D. Biggs (Eds.), *Seals and Sealing in the Ancient Near East*: 89–105. Malibu, CA: Undena.

Larsen, M. T. 1987. Commercial networks in the ancient Near East. In M. Rowlands, M. T. Larsen, and K. Kristiansen (Eds.), *Centre and Periphery in the Ancient World*: 47–56. Cambridge: Cambridge University Press.

Leemans, W. F. 1960. *Foreign Trade in the Old Babylonian Period as Revealed by Texts from Southern Mesopotamia*. Leiden, Netherlands: E. J. Brill.

Lehrberger, G. 1995. The gold deposits of Europe. An overview of possible metal sources for prehistoric gold objects. In G. Morteani and J. P. Northover (Eds.), *Prehistoric Gold in Europe: Mines, Metallurgy, and Manufacture*: 115–44. Dordrecht, Netherlands: Kluwer.

Lichtheim, M. 1975. *Ancient Egyptian Literature I. The Old and Middle Kingdoms*. Berkeley: University of California Press.

Limet, H. 1985. Les Relations entre Mari et la Côte Mediterranéenne sous le Regne de Zimri-Lim. In E. Gubel and E. Lipinski (Eds.), *Phoenicia and Its Neighbours*: 13–20. Leuven, Belgium: Peeters.

Lipinski, E. 1988. The socio-economic condition of the clergy in the Kingdom of Ugarit. In M. Heltzer and E. Lipinski (Eds.), *Society and Economy in the Eastern Mediterranean, c. 1500–1000 B.C.*: 125–50. Leuven, Belgium: Peeters.

Liverani, M. 1979. "Irrational" elements in the Amarna trade. In M. L. Jaffe (Ed.), *Three Amarna Essays, Number 1 in Monographs on the Ancient Near East*: 93–106. Malibu, CA: Undena.

Liverani, M. 2000. The Great Powers club. In R. Cohen and R. Westbrook (Eds.), *Amarna Diplomacy. The Beginnings of International Relations*: 15–27. Baltimore: John Hopkins University Press.

Liverani, M. 2003. The influence of political institutions on trade in the Ancient Near East (Late Bronze Age to Early Iron Age). In C. Zaccagnini (Ed.), *Mercanti e Politica nel Mondo Antico*: 119–37. Rome: Bretschneider.

Lockhart, H. E. 1997. A paradigm for packaging. *Packages Technology and Science* 10:237–52.

Manning, P. 2008. Barista rants about stupid customers at Starbucks: What imaginary conversations can teach us about real ones. *Language and Communication* 28:101–26.

Marcus, E. 2002. Early seafaring and maritime activity in the southern Levant from prehistory through the Third Millennium BC. In E. C. M. van den Brink and T. E. Levy (Eds.), *Egypt and the Levant: Interrelations from the 4th through the early 3rd Millennium* BC: 403–17. London: Leicester University Press.

Mauss, M. and E. Durkheim. 1963. *Primitive Classification*. London: Cohen and West.

McGovern, P. E. 1997. Wine of Egypt's golden age: An archaeochemical perspective. *Journal of Egyptian Archaeology* 83:69–108.

McGovern, P. E. 2003. *Ancient Wine. The Search for the Origins of Viniculture*. Princeton, NJ: Princeton University Press.

McGraw, A. P. and P. E. Tetlock. 2005. Taboo trade-offs, relational framing, and the acceptability of exchanges. *Journal of Consumer Psychology* 15:2–15.

McGraw, A. P., P. E. Tetlock, and O. V. Kristel. 2003. The limits of fungibility: Relational schemata and the value of things. *Journal of Consumer Research* 30:219–29.

Meeks, D. 1993. La production de l'huile et du vin dans l'Egypte Pharaonique. In M. C. Amouretti and J. P. Brun (Eds.), *La Production du Vin et de l'Huile en Mediterranée*: 3–38. Athens: Ecole Française d'Athènes (Bulletin de Correspondences Hélleniques Supplement 26).

Melena, J. 1983. Olive oil and other sorts of oil in the Mycenaean Tablets. *Minos* 18:89–123.

Meneley, A. 2007. Like an extra virgin. *American Anthropologist* 109:678–87.

Merkel, J. 1986. Ancient smelting and casting of copper "oxhide" ingots. In M. S. Balmuth (Ed.), *Studies in Sardinian Archaeology II: Sardinia in the Mediterranean*: 251–64. Ann Arbor: University of Michigan Press.

Merrillees, R. S. 1962. Opium trade in the Bronze Age Levant. *Antiquity* 36:287–92.

Michailidou, A. 1999. Systems of weight and social relations of "private" production in the late Bronze Age. In A. Chaniotis (Ed.), *From Minoan Farmers to Roman Traders. Sidelights on the Economy of Ancient Crete*: 87–113. Stuttgart, Germany: Franz Steiner Verlag.

Michailidou, A. 2001. Recording quantities of metal in Bronze Age societies in the Aegean and Near East. In A. Michailidou (Ed.), *Manufacture and Measurement. Counting, Measuring and Recording Craft Items in Early Aegean Societies*: 85–120. Athens: Boccard.

Miller, D. 2008. The uses of value. *Geoforum* 39:1122–32.

Molm, L. D., G. Peterson, and N. Takahashi. 2001. The value of exchange. *Social Forces* 80:159–84.

Moorey, P. R. S. 1994. *Ancient Mesopotamian Materials and Industries*. Oxford: Oxford University Press.

Moran, W. L. 1987. *The Amarna Letters*. Baltimore: John Hopkins University Press.

Muhly, J. D. 1996. The significance of metals in the late Bronze Age economy of Cyprus. In V. Karageorghis and D. Michaelides (Eds.), *The Development of the Cypriot Economy. From the Prehistoric Period to the Present Day*: 44–59. Nicosia: University of Cyprus.

Muniz, A. and T. C. O'Guinn. 2001. Brand communities. *Journal of Consumer Research* 27:412–32.

Negus, K. 2002. The work of cultural intermediaries and the enduring distance between production and consumption. *Cultural Studies* 16:501–15.

Nougayrol, J., E. Laroche, C. Virolleaud, and C. F. A. Schaeffer. 1968. *Ugaritica V. Nouveaux Textes Accadiens, Hourrites et Ugaritiques des Archives et Bibliothèques Privées d'Ugarit*. Paris: Institut Français d'Archéologie de Beyrouth.

Oppenheim, A. L. 1964. *Ancient Mesopotamia: Portrait of a Dead Civilization*. Chicago: University of Chicago Press.

Oppenheim, A. L., R. H. Brill, D. Barag, and A. von Saldern. 1970. *Glass and Glassmaking in Ancient Mesopotamia*. Corning, NY: Corning Museum of Glass Press.

Oswald, L. R. 1999. Culture swapping: Consumption and the ethnogenesis of middle-class Haitian immigrants. *Journal of Consumer Research* 25:303–18.

Özgüç, N. and O. Tunca. 2001. *Kültepe-Kanis: Sealed and Inscribed Clay Bullae*. Ankara: Turkish Historical Society.

Palaima, T. 1984. Inscribed stirrup jars and regionalism in Linear B Crete. *Studi Micenei ed Egeo-Anatolici* 25:187–203.

Palaima, T. 1991. Maritime matters in the Linear B Tablets. In R. Laffineur (Ed.), *Thalassa. L'Egée Préhistorique et la Mer*: 273–310. Liège, Belgium: University of Liège.

Palmer, R. 1994. *Wine in the Mycenaean Palace Economy*. Liège, Belgium: University of Liège.

Palmer, R. 2000. Wine and viticulture in the Linear A and B texts of the Bronze Age Aegean. In P. E. McGovern, S. J. Fleming, and S. H. Katz (Eds.), *The Origins and Ancient History of Wine*: 269–85. London: Routledge.

Pernicka, E., C. Eibner, O. Oztunal, and G. A. Wagner. 2003. Early Bronze Age metallurgy in the North-East Aegean. In G. A. Wagner, E. Pernicka, and H.-P. Uerpmann (Eds.), *Troia and the Troad. Scientific Approaches*: 143–72. Berlin: Springer.

Pinker, S., M. A. Nowak, and J. J. Lee. 2008. The logic of indirect speech. *Proceedings of the National Academy of Sciences* 105:833–38.

Polanyi, K. 1957. Marketless trading in Hammurabi's time. In K. Polanyi, C. Arensberg and H. Pearson (Eds.), *Trade and Markets in the Early Empires*: 12–26. Glencoe, IL: Free Press.

Postgate, J. N. 2001. System and style in three Near Eastern bureaucracies. In S. Voutsaki and J. Killen (Eds.), *Economy and Politics in the Mycenaean Palace States*: 181–94. Cambridge: Cambridge Philological Society.

Powell, M. A. 1995. Wine and the vine in ancient Mesopotamia. In P. E. McGovern, S. J. Fleming, and S. H. Katz (Eds.), *The Origins and Ancient History of Wine*: 97–122. Luxembourg: Gordon and Breach.

Powell, M. A. 1999. Monies, motives and methods in Babylonian Economics. In J. G. Dercksen (Ed.), *Trade and Finance in Ancient Mesopotamia*: 5–23. Leiden, Netherlands: Nederlands Instituut voor het Nabije Oosten.

Pritchard, J. B. 1969. *Ancient Near Eastern Texts Relating to the Old Testament*. Princeton, NJ: Princeton University Press.

Pulak, C. 1997. The *Uluburun* shipwreck. In S. Swiny, R. L. Hohlfelder, and H. Wylde Swiny (Eds.), *Res Maritimae: Cyprus and the Eastern Mediterranean from Prehistory to Late Antiquity*: 233–62. Atlanta: Cyprus and American Archaeological Research Institute.

Pulak, C. 1998. The *Uluburun* shipwreck: An overview. *International Journal of Nautical Archaeology* 27:188–224.

Pulak, C. 2000. The balance weights from the Late Bronze Age shipwreck at Uluburun. In C. F. E. Pare (Ed.), *Metals Make the World Go Round. The Supply and Circulation of Metals in Bronze Age Europe*: 247–66. Oxford: Oxbow.

Rahmstorf, L. 2006. Zur Ausbreitung vorderasiatischer Innovationen in die frühbronzezeitliche Agäis. *Praehistorische Zeitschrift* 81:49–96.

Renfrew, A. C. 1986. Varna and the emergence of wealth in later prehistoric Europe. In A. Appadurai (Ed.), *The Social Life of Things*: 141–68. Cambridge: Cambridge University Press.

Ribichini, S. and P. Xella. 1985. *La Terminologia dei Tessili nei Testi di Ugarit*. Rome: Consiglio Nazionale delle Ricerche.

Richardson, G. 2008. Brand names before the Industrial Revolution. *National Bureau of Economic Research Working Papers* (13930).

Roth, A. M. 1992. The pss-kf and the "opening of the mouth" ceremony. *Journal of Egyptian Archaeology* 78:113–47.

Säve-Söderbergh, T. 1957. *Four Eighteenth Dynasty Tombs*. Oxford: Oxford University Press.

Schaeffer, C. F. A. 1949. *Ugaritica II (Missions de Ras Shamra V)*. Paris: C. Klincksieck.

Schallin, A.-L. 1997. The late Bronze Age potter's workshop at Mastos in the Berbati Valley. In C. Gillis, C. Risberg, and B. Sjöberg (Eds.), *Trade and Production in Premonetary Greece: Production and the Craftsman*: 73–88. Jonsered, Sweden: Paul Aström.

Serpico, M. 2004. Natural product technology in New Kingdom Egypt. In J. Bourriau and J. Phillips (Eds.), *Invention and Innovation. The Social Context of Technological Change 2. Egypt, the Aegean and the Near East 1650–1150 BC*: 96–120. Oxford: Oxbow.

Serpico, M. and R. White. 2000. Oils, fat and wax. In P. T. Nicholson and I. Shaw (Eds.), *Ancient Egyptian Materials and Technologies*: 390–429. Cambridge: Cambridge University Press.

Shelmerdine, C. W. 1985. *The Perfume Industry at Mycenean Pylos*. Gothenburg, Sweden: Paul Aström.

Shelmerdine, C. W. 1995. Shining and fragrant cloth in Homer. In J. Carter and S. Morris (Eds.), *The Ages of Homer. A Tribute to Emily T. Vermeule*: 99–107. Austin: University of Texas Press.

Shelmerdine, C. W. and J. Bennet. 2008. Mycenaean states. Economy and administration. In C. W. Shelmerdine (Ed.), *The Cambridge Companion to the Aegean Bronze Age*: 289–309. Cambridge: Cambridge University Press.

Shennan, S. J. 1999. Cost, benefit and value in the organisation of early European copper production. *Antiquity* 73:352–62.

Shennan, S. J. 2002. *Genes, Memes and Human History*. London: Thames and Hudson.

Sherratt, A. G. 1993. What would a Bronze Age world-system look like? Relations between temperate Europe and the Mediterranean in later prehistory. *Journal of European Archaeology* 1:1–57.

Sherratt, A. G. and E. S. Sherratt. 1991. From luxuries to commodities: The nature of Bronze Age trading systems. In N. Gale (Ed.), *Bronze Age Trade in the Mediterranean*: 351–81. Gothenburg, Sweden: Paul Aström.

Sherratt, E. S. 1982. Patterns of contact: Manufacture and distribution of Mycenean pottery 1400–1100 BC. In J. G. P. Best and N. M. W. de Vries (Eds.), *Interaction and Acculturation in the Mediterranean*: 179–95. Amsterdam: B. R. Gruener.

Sherratt, E. S. 1994. Commerce, iron and ideology: Metallurgical innovation in 12th–11th century Cyprus. In V. Karageorghis (Ed.), *Cyprus in the 11th Century* BC: 59–106. Nicosia, Cyprus: Leventis Foundation.

Sherratt, E. S. 1999. E pur si muove: Pots, markets and values in the second millennium Mediterranean. In J. P. Crielaard, V. Stissi, and G. J. van Wijngaarden (Eds.), *The Complex Past of Pottery: Production, Circulation and Consumption of Mycenean and Greek Pottery (Sixteenth to Early Fifth Centuries* BC*)*: 163–209. Amsterdam: Gieben.

Sherratt, E. S. 2000. Circulation of metals and the end of the Bronze Age in the Eastern Mediterranean. In C. F. E. Pare (Ed.), *Metals Make the World Go Round. The Supply and Circulation of Metals in Bronze Age Europe*: 82–98. Oxford: Oxbow.

Shimy, M.A.-H. 1997. *Parfums et Parfumerie dans l'Ancienne Egypte (de l'Ancien Empire à la Fin du Novel Empire)*. Villeneuve d'Asq, France: Septentrion.

Silver, M. 1995. *Economic Structures of Antiquity*. Westport, CT: Greenwood Press.

Silverstein, M. 2006. Old wine, new ethnographic lexicography. *Annual Review of Anthropology* 35:481–96.

Simmel, G. 1900. A chapter in the philosophy of value. *American Journal of Sociology* 5:179–86.

Smith, L. M. V., J. D. Bourriau, Y. Goren, M. J. Hughes, and M. Serpico. 2004. The provenance of Canaanite amphorae found at Memphis and Amarna in the New Kingdom: Results 2000–2002. In J. Bourriau and J. Phillips (Eds.), *Invention and Innovation. The Social Context of Technological Change 2. Egypt, the Aegean and the Near East 1650–1150* BC: 55–77. Oxford: Oxbow.

Stos-Gale, Z. A., G. Maliotis, N. H. Gale, and N. Annetts. 1997. Lead isotope characteristics of the Cyprus ore deposits applied to the provenance studies of copper oxhide ingots. *Archaeometry* 39:83–124.

Tallet, P. 1998. Quelques aspects de l'economie du vin en Egypte ancienne, au Nouvel Empire. In N. Grimal and B. Menu (Eds.), *Le Commerce dans l'Egypte Ancienne*: 240–67. Paris: L'Institut Français d'Archéologie Orientale.

Tallet, P. 2004. Les oliveraies du Pharaon au Nouvel Empire. In B. Menu (Ed.), *La Dépendance Rurale dans l'Antiquite Egyptienne et Proche-Orientale*: 53–75. Paris: L'Institut Francais d'Archeologie Orientale.

Tournavitou, I. 1995. *The "Ivory Houses" at Mycenae*. London: British School at Athens.

Tuk, M. A., P. W. J. Verlegh, A. Smidts, and D. H. J. Wigboldus. 2009. Sales and sincerity: The role of relational framing in word-of-mouth marketing. *Journal of Consumer Psychology* 19:38–47.

Twede, D. 2002. Commercial amphoras: The earliest consumer packages? *Journal of Macromarketing* 22:98–108.

Tzedakis, Y. 1969. L'atelier de ceramique Postpalatial à Kydonia. *Bulletin de Correspondences Helléniques* 93:396–418.

van Alfen, P. 1997. The LM IIIB inscribed stirrup-jars as links in an administrative chain. *Minos* 31–32:251–74.

van der Toorn, K. 1996. *Family Religion in Babylonia, Syria and Israel*. Leiden, Netherlands: E.J. Brill.

van Soldt, W. H. 1990. Fabrics and dyes at Ugarit. *Ugarit-Forschungen* 22:321–57.

van't Hooft, P. P. M., M. J. Raven, E. H. C. van Rooij, and G. M. Vogelsang-Eastwood. 1994. *Pharaonic and Early Medieval Egyptian Textiles*. Leiden, Netherlands: Rijksmuseum van Oudheden.

Veenhof, K. R. 1972. *Aspects of Old Assyrian Trade and Its Terminology*. Leiden, Netherlands: Brill.

Veenhof, K. R. 2003. Trade and politics in ancient Assur. Balancing of public, colonial and entrepreneurial interests. In C. Zaccagnini (Ed.), *Mercanti e Politica nel Mondo Antico*: 69–118. Rome: Bretschneider.

Veenhof, K. R. and J. Eidem. 2008. *Mesopotamia. The Old Assyrian Period*. Freiburg, Germany: Academic Press.

Ventris, M. and J. Chadwick. 1973. *Documents in Mycenaean Greek*. Cambridge: Cambridge University Press.

Vercoutter, J. 1959. The gold of Kush. *Kush* 7:120–53.

Vogelsang-Eastwood, G. 2000. Textiles. In P. T. Nicholson and I. Shaw (Eds.), *Ancient Egyptian Materials and Technology*: 268–98. Cambridge: Cambridge University Press.

Vroom, J. 2003. *After Antiquity. Ceramics and Society in the Aegean from the 7th to the 20th Century A.C. A Case Study from Boeotia, Central Greece*. Leiden, Netherlands: University of Leiden Faculty of Archaeology.

Wachsmann, S. 1998. *Seagoing Ships and Seamanship in the Bronze Age Levant*. London: Chatham.

Wagner, G. A., G. Weisgerber, and W. Kroker. 1985. *Silber, Blei und Gold auf Sifnos: prähistorische und antike Metallproduktion*. Bochum, Germany: Deutsches Bergbau-Museum.

Wallendorf, M. 2001. Literally literacy. *Journal of Consumer Research* 27:505–11.

Wallerstein, I. 1974. *The Modern World-System: Capitalist Agriculture and the Origins of the European World-Economy in the Sixteenth Century*. New York: Academic Press.

Watson, W. G. E. 1999. Wonderful wine (KTU I.22: 17-20). *Ugarit-Forschungen* 31:776–84.

Watts, D. J. and S. H. Strogatz. 1998. Collective dynamics of "small-world" networks. *Nature* 393:440–42.

Webb, J. M. 2002. Device, image, and coercion: The role of glyptic in the political economy of late Bronze Age Cyprus. In J. S. Smith (Ed.), *Script and Seal Use on Cyprus in the Bronze and Iron Ages*: 111–54. Boston: Archaeological Institute of America.

Weeks, L. R. 2003. *Early Metallurgy of the Persian Gulf. Technology Trade, and the Bronze Age World*. Leiden, Netherlands: Brill.

Wengrow, D. 2008. Prehistories of commodity branding. *Current Anthropology* 49:7–34.

Wengrow, D. forthcoming. "Fleshpots of Egypt": Rethinking "temple economy" in the Bronze Age Mediterranean. In E. Frood, A. McDonald, and R. Parkinson (Eds.), *Essays for John Baines*. Oxford: Griffith Institute.

Whitehead, H. 1993. Morals, models and motives in a different light: A rumination on Alan P. Fiske's *Structures of Social Life*. *Ethos* 21:319–56.

Whitelaw, T. 2001. Reading between the tablets: Assessing Mycenaean palatial involvement in ceramic production and consumption. In S. Voutsaki and J. Killen (Eds.), *Economy and Politics in the Mycenaean Palace States*: 51–80. Cambridge: Cambridge Philological Society.

Wilk, R. R. 2006. *Home Cooking in the Global Village: Caribbean Food from Buccaneers to Ecotourists*. Oxford: Berg.

Winlock, H. E. 1945. *Slain Soldiers from the Tomb of Nebhepetre Mentuhotpe*. New York: Metropolitan Museum of Art.

Wiseman, D. J. 1953. *The Alalakh Tablets*. Liverpool, UK: Tinling.

Yener, K. A., P. B. Vandiver, and L. Willies. 1993. Reply to J. D. Muhly, "Early Bronze Age Tin and the Taurus." *American Journal of Archaeology* 97:255–64.

Zaccagnini, C. 1984. Transfers of movable property in Nuzi private transactions. In A. Archi (Ed.), *Circulation of Goods in Non-Palatial Context in the Ancient Near East*: 139–60. Rome: Edizioni dell'Atene.

Zaccagnini, C. 1987. Aspects of ceremonial exchange in the Near East during the Late Second Millennium B.C. In M. Rowlands, M. M. Larsen, and K. Kristiansen (Eds.), *Centre and Periphery in the Ancient World*: 57–65. Chicago: Oriental Institute.

Chapter 3

The Work of an Istanbulite *Imitasyoncu*

Magdalena Crăciun[1]

A counterfeiter illegally uses a company's well-established logo, packing, and reputation for his own benefit. However, I argue in this chapter that a counterfeiter might be regarded as an economic and cultural agent, actively shaping, transforming, qualifying, and mediating brands. He is an anonymous contributor to the collaborative construction of a brand (Wilk 2006), alongside recognised actors such as marketers, advertisers, retailers, or celebrities. Being located in the underbelly of what is publicly acknowledged, supporting, nevertheless, the visible and the acknowledged, his work falls outside the usual (and legal) definition of a cultural intermediary (Bourdieu 1984; Negus 2002) and remains invisible as a result.

This chapter addresses itself to this neglected contribution and details the work of an Istanbulite *imitasyoncu*, that is, a maker and seller of imitations, who has managed, despite the vagaries of the social, economic, and legal environment, to turn the counterfeiting of brands into a profession and to gain a relatively stable position on the local market. I describe the routines of such a business, pointing out how this man mediates various interests and tastes in the process of choosing brands to imitate and models to mutiply, dealing with manufacturers, suppliers, and buyers and making sense of his customers. His success depends on having the appropriate cultural capital, in the form of knowledge of fashion trends, brands, and names in the business at the local level.

In the first part of the chapter, I place this choice of activity within the main character's life story, suggesting that, in this case, the fake brand can be understood, following Bourdieu (1977), as the principal locus for the objectification of the structuring principle that governs the *imitasyoncu*'s life. Furthermore, this choice is contextualised within the Istanbul clothing industry. In the second part, I focus on everyday life: from the way the *imitasyoncu* obtains the goods in which he is trading, to the array of commodities in his shop, to the everyday activities of conducting a fluid business, infringing the law, and getting on with the work.

I will call this Istanbulite trader Fırlama, using one of his old nicknames, which is given usually to bold, impatient, practical-minded children who demonstrate a keen sense for social manoeuvring at an early age. He has allowed me, an anthropologist interested in fake brands, to drop by anytime I want and to poke my nose into his business and his life. After interviews with people vigilant about sharing information about their illegal occupation, I was looking for someone willing to share the juicy stuff I wanted to know about: the daily practice of dealing in fake branded clothing in Istanbul. Fırlama, slowly recovering from his most severe blow, a lawsuit that forced him to hide his merchandise and reduce the pace of his business, gradually welcomed the anthropologist who was interested in listening to his ideas about this kind of business and its prosecution as well as the opportunity it presented to him for self-reflection and self-presentation. Our encounter can be thus described as an instance of the "complicity of mutual interest between anthropologist and informant, subtly but clearly understood by each other, that makes rapport possible—indeed that constitutes, even constructs it" (Marcus 1997: 89).

Choosing Imitations

Fırlama started his life under very strict rules, but responded by systematically rejecting them; he needed illegality to confirm himself as an individual rather than the mere product of these rules. His father was a conservative trader who tried to make his children dependent on his wisdom, experience, and judgement. The boys came to work in the bazaar about the age of 5, for their father decided they had to learn quickly how to earn a living. For his youngest son, Fırlama, he had a different plan, wanting him to be an imam (Muslim religious leader). The boy spent 4 years in an imam-*hatip* (preacher-prayer leader) school, an institution characterised by an atmosphere of discipline, where the duty of the students is to obey and the task of the teachers is to inculcate moral values and compel students to be adherents to and practitioners of Muslim teachings (Pak 2004). Upon finishing this school, Fırlama rebelled and left home. For years, he lived a colourful life on the streets, had all manner of experiences, and had made the most of doing everything that was previously forbidden.

However, Fırlama had only one wish for life, namely to prove himself. The army, one of the important institutions responsible for the production of masculinity in Turkey (Kandiyoti 1994; Neyzi 2001), played a major role in crystallising his personal habitus. Demonstrating resourcefulness, wit, and bravery, Fırlama began to value his difference and to search for ways of combining his predilection for adventure with modalities of earning a living and situating oneself in society. As he recounts:

> My father put the label on my forehead. He said I was a loser and nothing good would come out of me. I wanted to prove myself to my family. … So I needed to make money regularly. I was used to making and trading in things that sell well anyway. My father did this. Leather was so much in demand in the '60s and '70s. In the '70s came foam rubber; cotton and wool were gone, foam rubber sold well, and my father was in this business. Some people say "It is ok if I don't earn a lot as long as I don't get into trouble." With me it's the opposite. I want to earn a lot but I don't care if I get into trouble. That is my character. Little by little, I came to realise smuggling, imitation are the kind of jobs I would make money everyday, even if they are dangerous ones.

Fırlama could have chosen to enter deeper into the underworld or to become a spy, as the officer he worked for during his military service encouraged him, but smuggling and faking seemed the best options, the "cleanest" ones.

Therefore, since the early 1980s, Fırlama has been involved in the trade in fake brands, fake-branded underwear in his case. "In those times," he recalls:

> Turkey was gradually opening to the world. Because workmanship was so cheap, smart merchants came and had their stuff produced here. Somebody asked for CK, for England, a big smuggling deal. Nobody knew of any law against it. And I did not even know that "CK" stood for Calvin Klein. We know it as Ce Ka but the Europeans say CK. It settled in my brain as Ce Ka. I had already the tendency towards such things, I dived right in. Still I have not surfaced.

Today, Fırlama occupies a middle-level position in the market, between the big players who invest millions in this business and the insignificant participants who stitch poorly cut labels on cheap clothes; in other words, he is between two different moral universes—large-scale, high-tech, high-income business and small-scale, low-tech, subsistence-level activities. After many years in this trade, he is an experienced actor, his current arrangement including a manufacturing site, partnerships with different workshops, a few secluded storehouses, and a shop.

A Favourable Context

Two concomitant processes favoured the profitable business of manufacturing and selling imitations, namely the transformation of Istanbul into a manufacturing site in the global clothing industry and the city's recapturing of its historical role as a regional market, with trading routes spreading across a vast area that includes eastern Europe and the former Soviet Union.

Since the 1980s, a significant number of Turkish firms have entered the web of the global clothing industry. Their entry has been favoured by a combination of factors: a national liberal policy of shifting to a market-oriented and an outward-looking strategy that encouraged buyer-driven industries; the arrival of European corporations that established plants in Turkey and of European designers who signed franchising and licensing agreements with local firms; the progressive removal of quotas on textile and clothing imports by the European Union in the 1990s; and the gradual lifting of U.S. quotas limiting imports of clothing from developing countries since the late 1980s, together with tariff cuts. Turkey has emerged as a low-cost producer, ranking ninth among the major clothing exporters of the world (WTO 2006) and manufacturing major global brands.

The clothing industry is "a cascade of operations, each of which can be, and in practice is at one time or another, separated off and subcontracted, creating almost a continuum of firms arranged in a hierarchy of skill, power, and profitability" (Peters et al. 2002: 229). Most of the Turkish contractors and subcontractors are located at the bottom of this hierarchy as simple firms that are responsible for sewing and assembly (highly labour-intensive tasks) of garments, their competitiveness lying in the relatively low wages and geographical proximity to the European Union. Some of them have upgraded their production, moving from assembly to full-package production, a move that brings the benefits of quality, flexibility, and productivity by forcing manufacturers to develop better technologies and stronger ties with local and foreign suppliers and subcontractors to satisfy the demands of marketers and retailers. This, in turn, leads to better motivation and performance, so that firms learn how to "do things better" and how to "make better things" (Schmitz and Knorringa 2000: 181).

Though manufacturers that operate in global networks face discouragement and even obstacles if they want to upgrade into higher value–added activities, such as design, brand-name manufacturing, marketing, and retailing, a few Turkish firms have successed in evolving to the point where they manufacture original brand names, Mavi Jeans being the most quoted example. Moreover, some large domestic manufacturers experienced a cautious and gradual transformation from industrial capital to commercial and financial capital (Tokatlı 2003).

A profitable side business has also proliferated in Turkey—that is, the trade in "slightly imperfect" items. Known as "export excess" products, these items may constitute up to 10 percent of a brand's production. As Tokatlı and Kızılgün (2004: 228) point out:

> This may or may not be considered part of the counterfeiting business. The "export excess" articles are not necessarily "fakes" in the sense of being copies of well-known brands, since they are manufactured at the same factories, together with the items that pass quality inspections, without any intent of counterfeiting. Rather, they represent a "loophole" in the networks, a source of fragility by which a significant accumulation of capital is possible by fullpackage manufacturers through the collection of unearned brand rents; [it] can be exploited by seemingly powerless members of the [global clothing industry] network.

Although it is impossible to know which firms take advantage of this trade and to what extent, the business seems to be profitable, for the excess products can be sold not only in the domestic market but also in the countries of the former Soviet Union and eastern Europe.

After the fall of communism and the reopening of the eastern European and Soviet borders, petty traders from all over this region have started to travel regularly to Turkey, mainly to Istanbul, to purchase moderate quantities of consumer goods, such as garments and leatherwear. Known as "suitcase trade" or "shuttle trade," this flourishing trade has been, for the most part, unregistered and evaded taxes and customs duties as states are either unwilling or unable to regulate it (Blacher 1996; Keyder 1999; Konstantinov 1996; Yükseker 2004). Turkey, for its part, has turned a blind eye to this "unexpected informal globalization" (Keyder 1999: 22) because it represented a source of much-needed foreign currency for the national economy. At a high point in the mid-1990s, suitcase trade exports from Turkey to the former Soviet Union were estimated at $9 billion annually (U.S. currency), a significant sum compared with Turkey's official exports, which ranged between $13 and $27 billion per annum over the decade (Yükseker 2004: 49). As the impact of brand names on the consumer societies of this region was conspicous, small-scale traders expressed their interest in branded goods. And Turkish entrepreneurs offered them a variety of products, ranging from "imperfect" brands to "perfect" fakes.

In brief, at manufacturing sites of the global clothing industry, strategies are invented to make brands escape systems of control, mutate, and proliferate, as local agents harness, appropriate, or divert the power of brand for their own benefit. In Istanbul, trade in imitations emerges out of opportunities such as growing manufacturing capacities, knowledge transfer, available technology, excess products, and leakages and an increasing number of customers who are demanding branded goods at affordable prices.

Doing Imitations

Fırlama owns a tiny shop in a bazaar to which crowds throng daily, drifting past the storefronts, peeping into the windows, entering the shops, turning everything upside down, but nonetheless buying. The majority of the goods in this shop are produced in a rented workshop equipped with high-tech automated Italian machines that speed up the production process and reduce wastage of materials. The workshop is located in a neighbourhood on the Anatolian side of Istanbul, and the production of microfibre seamless "tubes" (soon-to-be imitation goods) takes place during the late hours, after "normal" production—subcontracted work from local clothing companies—is finished. As Fırlama puts it: "At five o'clock, the CDs are changed and the production starts. It is the imitations' time! There are no workers in the workshop, only engineers who know how to handle the machines. The rest of the people working there just take care of the products, packed them for their next destination." Fırlama rarely needs to go to the workshop, as the phone is sufficient for keeping in touch with his trustworthy man, to supervise the movement of the goods and to place new orders. This man comes every week to the shop and collects the wages of the employers and money for materials and reports on that week's situation. Fırlama runs the workshop together with a partner, the partnership being rather a "marriage of convenience"; the two keep an eye on each other—one day kisses, another day curses. As Fırlama explains:

> We both use checkbooks. We do not want to have a bad reputation among banks. If you mess up with one check, then you are doomed! One factory can make up to 100,000 items a month. If I can't sell all of it or sell to wrong people, there will be trouble. It's all risky. But with 50,000, if I can't sell them, I still have that amount of money in the safe box so there will be no problem. You can use checks for 30 years but when you cannot pay only one, you are finished! Then you are not welcome at the bank, you are not a *beyefendi* [gentleman] anymore!

Though tense, this is a fruitful relationship, with the two taking great care to target different customers and support each other when one of them does not have enough money to pay his part of the factory's expenses.

Fırlama has his own arrangements with other workshops in which the different operations necessary for the finishing of his products can be carried out. All the places through which his goods are moving, including storehouses and the bazaar, are strategically placed in various parts of the city. However, from a different point of view, this is a normal production chain in the local clothing industry. From the rented workshop, the tubes are sent to be dyed in a different location, in another neighbourhood on the Anatolian side. From there, goods are later moved to another site, in the same neighbourhood, a place where they eventually become imitations, with the brand names printed on them. Long-lasting Italian paint is used for imprinting, which Fırlama buys at a good price by first sending it to Syria and then routing it to Istanbul through smuggling networks. Once the items are painted and imprinted with brand names, the last stop is a workshop on the European side of Istanbul, where the tubes are shaped into men's and women's underwear. Cardboard boxes are produced in a different place. A part of these products are sent to the shop, another directly to customers, and the rest to secure storehouses. Fırlama changes their locations at least once a year to avoid perilous encounters.

The finished products are microfibre seamless underwear for men and women in different colours, a far greater variety than could be found in any official shop. In style, they range from men's boxer briefs with a thin or wide waistband, standard and hipsters' briefs, women's boxer briefs, g-string thongs, and hipsters' briefs. There are also fancier models produced mainly for buyers who cater for more sophisticated customers. At a first glance, the only difference seems to be the brand name. However, sharp eyes will spot the detailing in the fabric that differentiates one brand from another; if not, Fırlama is always keen to point out the differences. Most of the goods are branded Armani and D&G, the safest brands, as these corporations, he insists, do not deploy lawyers to protect their interests from imitators. This is described differently to the customers, who are told that these two brands are the most in-demand on the market. Moreover, other producers and retailers of imitations I talked to share the same opinion, their main arguments being the quantity and variety of imitations of these brands on the local market. However, lawyers and local associations (e.g., United Brands/Birleşmiş Markalar Derneği) claim exactly the opposite and give examples of raids to protect these brands. In smaller quantities, Fırlama has Calvin Klein, Ralph Laurent, Björn Borg, Dsquared, Gucci, and Versace, and, under the counter, Hugo Boss and Diesel, the most dangerous brands these days, with fierce lawyers protecting their interests.

Fırlama confesses to me that what he likes the most about this trade sector is that:

> I am never left with goods that were not sold. There is no such thing as out-of-fashion here. Let's say D&G is very fashionable in Germany this year. In a year it becomes fashionable in eastern Europe. In two years in Turkey, in four years in Azerbaijan. It goes like this, step by step. There is a brand whose name begins with a D, I do not remember now the name. I sold a lot of those this year,

> but I have 2000 left now in the storage. I am not worried because I know a Romanian or Yugoslavian customer will ask for that product in a year or two. Or it will be very popular in Turkey. Ten years ago nobody cared for D&G.

Fırlama also keeps original Turkish brands in his shop to cater for local tastes, as he himself, bazaar traders, and some customers are very fond of these kinds of underwear and socks. Nevertheless, to his surprise, brand has an impact even on these goods. As he puts it:

> Sometimes I do not understand people. We know each other for so many years. They know what kind of person I am. They come and see Öztaş cotton boxer briefs and want to buy. I tell them it is not 100% cotton, I know this very well, a friend of mine manufactured them. I tell them buy this or that, it might not be Öztaş, but it is quality. They are deaf. They want Öztaş. They want a brand. As if wearing branded boxers or non-branded boxers is not the same thing. As if they do not serve the same purpose, that is, to keep your buttocks. I sell brands, but sometimes I do not understand what are they good for.

Fırlama's reaction is that of the producer who knows the quality. Theirs is the reaction of the customers, for whom the brand guarantees the quality. But there are other times in which, for example, Fırlama takes pride in wearing an original Quicksilver t-shirt. He is a consumer, for whom "one voice of the brand is always information and the implied warranty of objective purity and safety [and] its other voice is persuasion" (Wilk 2006: 26).

Thanks to his numerous connections among clothing manufacturers, Fırlama also trades in imperfect items (*defolu*) and nonbranded goods that he transforms into fakes by imprinting brand names or sewing labels and logos on them. Though a great deal of information about such opportunities can be learned in face-to-face meetings or by phone, he employs an experienced older man to comb the clothing manufacturing districts of the city in search of quality products that can be routed through his business. Fırlama rarely needs to go in person to settle a deal that this man traces down, for, having been around for so many years, people know him, put stock in his words and are prepared to do business with his employee. This man gives periodical reports by phone, keeping Fırlama well informed about novelties on the market and buys in his name whatever he knows will meet his employer's standards, be it thread, paint, fabric, labels, or finished products. He drops by the shop to bring samples only when he needs Fırlama's opinion and his knowledge of fashion trends. Samples are carefully scrutinised with expert eyes that know good trimmings and seams and experienced hands that can evaluate the fabrics. Though Fırlama does believe in and benefit from this man's ability to discern good quality things, in the end, it is only Fırlama who decides which products are marketable and which have the potential to become convincing fakes that can be displayed in his shop next to the imitations he manufactures.

Fırlama shows me some of this man's discoveries. One example is a small series of slightly imperfect white and black g-string thongs, of a better quality than those he can manufacture, on whose waistband he imprinted Armani and D&G. Several times, I witness him claiming these are original items and then regretting having done this, not only because he is not believed and the customers buy nothing in the end, but also because he realises that he is thereby admitting that his other products are of an inferior quality.

Some white camisoles are another lucky find. Soon after they arrived in the shop, he starts to talk them up, telling everyone they are identical to a model from the brand's next season collection. He puts one item on the table, takes a tape measure, and begins the demonstration:

> The margin, 2 centimetres. The rest, 47 centimetres, exactly as the original. Then there are the colours: blue as blue as Armani's blue, white as white as Armani's white, black as black as Armani's black, cyclamen more beautiful than Armani's cyclamen. And the brand name imprinted with Italian paint. They will sell like hot cakes, be sure of this, my friend.

He tells this to every visitor who crosses the shop's threshold the week he received the camisoles. When they are no longer new arrivals and not selling like hot cakes, he adds matching boxer briefs and advertises them as a set. A part of the cyclamen stock is redyed black and the brand name is changed to D&G. Long after, when I help him arrange the shelves, he recounts how his man found them on the market, finished white products, and how they both liked them and decided to turn them into fakes. The 2 and 47 centimetres story I once heard was invented on the spot in an attempt to convince his visitor to buy the merchandise.

Fırlama himself is a buyer of cotton underwear, working with several manufacturers at the same time, accepting their products, or demanding models that he knows he can sell for good profit through his own network of buyers and customers. From one of his partners, for example, he buys stretch cotton boxer briefs with a thin waistband. At the beginning, the items were of excellent quality, extra stock of an outdate model that a foreign buyer had refused. Fırlama bought that stock, sold it very well, and decided to work again with that manufacturer. As usual, he first discussed the model with his young apprentice. The boy is fond of trying all these new products and feeds back his experience and details about their fit on the body. And, as the boy likes to show off in front of his friends, he can report on their reactions. Fırlama and his apprentice agreed to change the original design slightly, deciding on another colour scheme, enlarging the front part, choosing smaller labels and more brand names, and also ordering button-fly boxer briefs in the same fabric. Since then, Fırlama's partner has been responsible for the whole manufacturing process, and as a result, waistbands are rougher and labels carelessly cut. Nevertheless, they sell well and bring good profit.

Fırlama and his apprentice constantly rearrange the shelves and the piles on the floor and fill up the empty spaces with patience and ingenuity. The front counter deserves special attention, for this is the first to fall victim to the eager hands of the customers. After every visit, the pairs of underwear are again neatly aligned and wrapped so that their brand remains visible. Huge bundles are set down by porters in front of the shop a few times per week and their content is quickly divided between the shop and the nearby storehouse. Thus, though crammed full with goods, the shop is always tidy and ready to be explored.

Trading in Imitations

Eager to sell his products, Fırlama waits for visitors: sitting on a stool at the entrance to the shop, smoking, sipping tea, talking with other merchants, eavesdropping on the ebb and flow of conversations, and scrutinising the bazaar alley. The earnest shopkeeper springs to his feet whenever customers approach, welcoming them with honeyed words, encouraging them to rummage through the shelves. "Look! Armani! Dolce&Gabbana!" some cry, thrilled. "Yes, sir, I have them all. In sixteen colours. Yes," Fırlama replies, waiting to take their order, tempting them with more colours than those displayed at the counter. Occasionally, he goes into details regarding the manufacturing, emphasising the excellent choice the customer is about to make, for his sells for €2 quality products that in the official shop costs €40. For all the customers, the prices are fixed, YTL 5 (Turkish currency, equal to $3 USC) a pair, YTL 6 a pair, depending on the day and the shopkeeper's mood. Nevertheless, Fırlama rarely forgets to emphasise that this is a special offer, "only for you, my friend, because I like you."

The bothersome customers, who imagine that they can get a discount by buying three pairs instead of two, are cut short and left alone to ponder whether they want to buy or not. "Bargaining, as if he buys a car," Fırlama usually comments, putting back on the shelves the things that seemed too expensive to his customer. The hesitaters are shown different items in the hope they would finally reach a decision, but many times their obvious pleasure in turning the shop upside down indisposes the shopkeeper. The few naifs, who think that they have made a lucky find, brands for a fraction of their price, are assured the goods are imitations, but quality products, hard to find in other bazaar shops unless the shopkeepers are his clients. During these encounters, without exception, there comes a moment when both parties stretch the underwear, Fırlama to demostrate their resistance and quality, the customers to check whether they suit them or not. Rarely will one leave with just one pair, for there is no harm in buying more, things being dirt cheap in this shop, people often think aloud. Fırlama invites those customers who buy from him for the first time to immediately wash and wear a pair to assess quality via consumption. They will like his products and will return to buy more, he is certain. "I am here. You can find me again," he says, instead of good bye.

Now and then agitated bazaar shopkeepers, who sell Fırlama's products at a mark-up, rush into the shop, in urgent need of articles in a certain colour, size, or brand. "I will take fifty pairs, white, black and grey, all sizes, all brands, for the whole week," one places the order, quite determined. "My friend, I need right now three black pairs, I have a client in my shop, it does not matter what brand, just hurry up," another utters at once, breathing heavily with effort, only to bring them back, few minutes later, apologising, again in one breath, because the client has changed his mind. "Give me the sexiest colours you have, brother, but only Armani," speaks another, in quite a humdrum manner. They are promptly served and the number of items they take is written on their page in Fırlama's notebook, the money being collected later, usually at the end of the week.

Fırlama is the supplier for tens of such small-scale buyers from the bazaar and its vicinity, but these partnerships are of short duration and most of the small-scale buyers are bad at settling their debts. These shopkeepers earn a living from selling goods with the highest possible profit to occasional customers. Forging long-lasting relationships with small-scale buyers and customers is not a thing they worry about. Therefore, they can quickly change their supplier, on condition that the products are cheaper. Taking this into consideration, Fırlama is open to working with any new shopkeeper who wants to buy his goods, provided that this man has someone to introduce him and speak on his behalf, confirming that he is trustworthy and that his shop is well positioned.

Important buyers who are not based in Istanbul place orders by phone, usually large ones, to be sent to nearby places, distant tourist resorts, or abroad. These are hard-earned and well-established relationships through which Fırlama sells a significant part of his merchandise. At these times, there is a lot of work to do. Fırlama summons his apprentice to help him arrange things. He likes to do such things as quickly as possible and, consequently, these are the times when the boy is scolded for failing to keep pace. Packets are bundled up, wrapped in plastic, fastened with rubber bandbands, and left in front of the shop, ready to be picked by porters from the cargo company Fırlama is working with. He gives the cargo company his full name and address, a habit many others consider foolish. Fırlama reasons that in case something bad does happen, everybody will become a stranger all of a sudden (*herkez Fransız oluyor*), denying all responsibility.

Besides customers and other shopkeepers, friends, trusthworthy partners, and serious buyers cross Fırlama's threshold. They are offered tea, a cigarette, and a chat, and they reject none. Injustices they have suffered, usually unpaid debts, are eagerly recounted, the men seething with indignation, uttering harsh but well-deserved words. Known as an honest person, Fırlama is the best audience, his most trenchant criticism being reserved for the fickle characters, the ignorants, and the liars. Doing business in Istanbul, especially doing this kind of business, means he needs to keep himself well informed about people and events, the latest scandals, and the hotest news. All the stories and all the people are so entangled in this world, such that "the liar's candle will not last forever," a way of saying that sooner or later wrongs and betrayals are discovered. In this way, participants distill pieces of noteworthy information from the gossip and allusions lavishly dispensed on these occasions.

Fırlama explains what these encounters are good for:

> Last month in Antalya, a guy went bankrupt. Ten minutes later, everyone in Istanbul found out. It works this way. If I cannot pay a check I have written, it will be heard all over Turkey in 10 minutes. Friends let you know, warn you to be careful when approached by a certain man. When a stranger comes here and says he wants to do business with me, I ask who else in the trade he knows, where is his shop, and then I check if he has betrayed other imitation sellers, if he pays regularly what he has to pay, if his shop is indeed at a nice place. If he is a trader who has been around, even if I do not know him, most probably a friend of mine will know him and tell me what kind of guy he is. My friend says, for example, "This one does not pay on time," so I am suddenly out of goods. Or "This one will cause no trouble," so it means I can trust him and behave accordingly. And this friend of mine will be my guarantor, he will have to pay in case the other does not, or, at least I will go to him and complain, threaten, discredit that guy. I do this. Whenever I have a problem, I let everyone know.

During these encounters, Fırlama does not miss a chance to talk up his goods.

Depending on the audience, his intention is either to convince them to buy or to consolidate his image as a successful trader and professional *imitasyoncu*. In those moments, as they say, his tongue is hung in the middle and wags at both ends. The high-tech factory, the clever way in which his business is organised, the quality of the products, the new models, the new brands, all eddy around in those torrents of words.

He presents the new arrivals, rendering their features as attractive and desirable, specifying their availability, and often referring to how well received they have been with other buyers and how crazy customers in this and that location are about them. He firmly believes that direct engagement with the products is the most convincing selling strategy, so he persuades his visitors to touch the fabrics and teaches them how quality feels. Sometimes he sends his apprentice to borrow a pair from another. The visitors are instructed in how to compare the different items and admit his products are better. In turn, the visitors want to know which goods are manufactured by him and which are not. The lastest rumour, they say, is that some traders are in fact dealing in goods made in China, but claiming that they are made in Turkey. As it is hard, if not impossible, to find out the manufacturing details of these goods, it is better to avoid them, some insist.

Fırlama assures his interlocutors all the goods in his shop, manufactured by him or not, are of good quality. Buying and selling quality is the most profitable business strategy, both parties often conclude. Rarely do these visitors need the goods to be identical imitations, reproducing in the smallest details their official counterparts. Fırlama cannot provide identical copies but he can offer quality copies in a wider range of colours than a brand's official production. Segmentation and targeting are thus also characteristics of the market of fake brands (Entwistle 2006).

Fırlama confidently declares that he designed a particular model, that he chose the colour combinations, and that the new arrival is "this season" or "next season." Hard as it is to decide if Fırlama is showing off or telling the truth, his visitors nonetheless cannot help but look at the items with renewed interest. Fırlama informs them which are the newest brands, names they have never heard of, but which will be soon demanded. Burberry, Dsquared, Ralph Lauren, or Björn Borg, he recites, with difficulty. They are kindly requested to keep him in mind as their possible trusthworthy supplier of these new brands. In brief, everything in this shop is special and whatever new deal the visitors and he decide on it is meant to be an unheard-of product and an irresistible offer. For Fırlama has the reputation of being a knowledgeable person, the one who can tell what is fashionable, what brands are demanded, who manufactures what, who has the technology for producing certain things, what to do if there is trouble with lawyers and police.

Samples are shown during these encounters, his or theirs. When offered samples, Fırlama evaluates their fabric, seams, and stitches and discuss their marketability. His own taste is critical to selectability. Someone shows him several pairs of stretch cotton boxer briefs, branded Ed Hardy and La Martina, for example. Though both brands sell well nowadays, Fırlama is not at all impressed. The logo is imprinted with quality paint and the fabric is of an acceptable quality, but the cut for him is entirely wrong, not fitted for a man's body. His partner did not want to listen to his advice and invested a considerable amount of money in merchandise cut this way and now they crowd his storehouse. Fırlama does not want to make the same mistake. Doing business means assuming risks but not making mistakes regarding the products, he reminds his interlocutors.

During these encounters, everyone has a successful story to share. Fırlama's most recent one is about Björn Borg. "Haven't you heard of this brand yet?" he interrogates his companion. A few pairs of original Björn Borg cotton boxer briefs remained in his shop for months until Fırlama made up his mind and bought the whole stock of a factory that was selling it on the local market after it was refused by a foreign buyer, a 1-week delay being the factory's only fault. Fırlama is now proud of being the only shopkeeper in the bazaar who can offer this brand and content to have many customers buying without thinking twice. Had these cotton boxer briefs in bright colours been produced by a Turkish company, he comments, nobody would even look at them. "Boxer shorts with flowers or flies," he repeats several times, scornfully. "It is the brand name that thrills everyone," he bitterly remarks.

To some of his guests, Fırlama reveals his newest and dearest plan. He wants to register his own brand, whose name will be *Here* and whose logo will be an arrow pointing down, a jest he hopes will lure people into buying his goods. Needless to say, he will continue to sell imitations, but under the counter, for the punishment is more severe in the case of an owner of a brand found to be selling fake branded goods.

Not all those entering Fırlama's shop are welcome, but he finds ways of taking advantage of these meetings, too. For even in a tense encounter there might be an opportunity for praising the goods. Even to a bad payer, turning up after months in which he did not answer the phone, new arrivals might be introduced. These men make such unexpected visits that the shopkeeper is left speechless for a moment. "Here you are," Fırlama huffs, a second later, eyes red with fury, and he briskly pulls out a notebook kept under a pile on the shelf. "A two-page-long debt," "a 1-year debt," he snaps at them and reads the dates at which such and such quantity was purchased. After harsh remarks, the matters are settled and the pages torn from the notebook.

Rarely do such men leave immediately. Usually they dare to ask what else they might be offered. Rapidly, Fırlama's face turns from sulk to cordiality and when he speaks again his voice does not have the same coarse tone as before. The newest and most expensive models are first shown, the classical models in daring colours follow and so on and so forth, for many a minutes. "How much is this?" "Does it sell well?" are the comment he usually elicits. There are cases in which his offers are refused, one by one. Alternatives are suggested, different shops, with cheaper stuff, YTL 2.5 instead of YTL 3.5, his price. However, Fırlama, annoyed again, emphasises that though seeming identical, those goods are polyester, in sharp contrast to his fine-quality things. "Force the fabric near the stitches and you will get a ladder in a few seconds," he frowns and purses his lips. "Besides, they cheat the customers, selling 10 pair packets minus one or two, all sizes packets minus one size. It is up to you, my brother. I can give you their business card, you can call them right now," he suggests. Upon their departure, Fırlama mutters under his breath: "Cheapness, this is what he wants, not imitations. At least, I got my money back." Some do return later and buy his merchandise. Others do not.

In some cases, solving financial issues necessitates more than the simple act of giving or receiving cash or cheques. Blandishments and appeals to every known social tie may be some of the alternative techniques, but equally possible are grovelling and haughtiness, threats and shouting. The end of the working week is particularly exhausting for Fırlama; there are numerous phone calls to be made to remind people that they owe him money, as he has to pay his employees and cover the workshop's expenses. In the morning, Fırlama consults his notebook and tells his apprentice from which shops in the bazaar he has to collect money. Chain-smoking as he tries to reach people by cell phone, he flies into a temper every time one promises he will pay a visit in half an hour and bring the debt, using the exact words as he did a week ago, a month ago, a year ago.

Next come the calculations, and Fırlama plunges into silence, a grumpy tradesman, evaluating the profit, planning the future movements, deciding what bills to pay and what purchases to make for the workshop, and then reconsidering everything and drawing up an alternative plan. "In this notebook of mine," he explains, "there are written YTL 140 000, all transactions depending on a word or a promise. There is no solid proof, no written contracts. It can't be written with imitations, of course. It is illegal." In this world, business propositions are carefully pondered over, for everyone has been cheated before and for everyone knows the local habit of bragging and promising, only to forget seconds later. Business itself is fluid and ever changing, requiring ongoing maintenance given that relationships with other producers, suppliers, and traders are easily lost, broken, and refashioned.

Illegal, but Moral

Bouncing back from his worst disasters, Fırlama has made an effort at carving out a place for himself in the marketplace. At the same time, he has built a reputation as an honest and reliable trader. As he explains to his chatterbox of an apprentice and me:

> I do not behave immorally and the customer prefers me. Buyers call and make an order, they do not ever come here. I have buyers from Antalya that I do not remember what they look like. I send the goods, they send my money. If I do not have the product they ask, I call, tell them and suggest an alternative. If they agree, we send them. For instance, there are now three boxes I prepared in my depot. The customer never came here. It is so difficult to come here from Macedonia. He sent me €3,600, he wants 2,000 items, he sent the list. I prepared exactly what is on the list, because I don't want to lose the trust of this guy. When the transportation is ready, he will tell me and I will send the goods. He is sure I will not cheat him, I am sure he will not forget me. It is as simple as this. If you do not have good intentions, you cannot be sure about the customer's constancy. If you are honest to the customer, he will like you even if he is from New Guinea!

Moreover, Fırlama feels bound to remind us that:

> The ones who live or work without discipline are destined to lose in the end. I was like that until 1994. I lost something like one million dollars and thought to myself if I had taken these precautions, this would not happen. After that I became more disciplined and earned the money I lost. I have certain rules. In my right pocket I have my money to spend but I never touch the money in my left pocket. I have my "death money" at home, I mean sickness, I can only spend that money in case of sickness. I have such strict rules. If my spending money in my right pocket is finished, I do not go anywhere. I go home and avoid spending anything. If I really have to buy something I use my credit card. I never touch my left pocket. If I did, my checks would bounce and I would get into debt. It is my shop's money.

Doing business requires careful thought and moderation, good cash flow, and respect for the customers. Forgetting these principles might result in losing everything. In their world, a bankrupt trader is a nobody. Ideally, Fırlama insists, the trader must put his whole heart into his work and must be honest and thrifty. Then he will prosper and his shop and storehouses will be full to the brim.

I was also assured that:

> Effort must be put into making frauds too. I make the products the way Mercedes does, I always make the material like the way it should be produced. For instance, I have made this product very nicely, if I do not have it sewed with mercerised fibre, it will be torn in less than 2 months. More, if I do not put this into water with the chemicals that soften the fabric, the cost will decrease a bit, but the number of my customers will also decrease. This product of mine has a wholesale price of €1.80. But in another place I can find you the same kind of boxer shorts for €1.30 or €1.60. Here it writes 90% microfibre and 10% elastan. We can take this to a textile lab and test it. If what is written is not true I am prepared to lose all my belongings in court. Or you can take the shorts and make me wear them on my head! I could make this 100% synthetic and write the same thing, and then I would sell it to you for €1.50. The only lie here is that it writes "made in Italy." Everything else is true. If I wrote "made in Turkey" I would not even have the chance to sell it. Then it would not be imitation. It would be pointless.

Selling quality imitations is what keeps Fırlama in business. "As long as your products are good, nobody can touch you in this business. Except for the lawyers, of course," he points out.

Fırlama was firstly caught in 2001, because of a ridiculous mistake. The number of the flat that he used as a storage place and the number of the building were the same, a pure coincidence. A team of lawyers was looking for the ground floor shop, to make a routine inspection. His employee happened to be smoking a cigarette in front of the entrance at the wrong time and candidly answered their question, indicating that number 56 was on the fifth floor. The lawyers entered paradise, a flat filled with fake branded goods, and Fırlama faced a heavy fine and years in jail. The punishment was, however, postponed for 5 years, for this is the period after which the files are cleared, a legal trick for whose application Fırlama generously bribed lawyers and judges. Consequently, he took more precautions, changing the location of his production sites more frequently, hiding the storehouses even better, arranging for the bank to sequester some of his properties, and strengthening his ties with local politicians, lawyers, and underworld leaders. Last time, Fırlama was caught in his shop, but kept himself out of further trouble by half-bribing and half-threatening the lawyer who paid the importunate visit.

Over the last few years, efforts to combat conterfeiting have intensified, but for someone like Fırlama, who lived a colourful life when he was young and who was imprisoned several times for smuggling,

> these problems with the law are like fairy tales for me. I do not care! It is a serious thing, I must admit, but more than anything else, it is comical. Pure comedy! The most important thing: you have to go to court clean-shaved. You have to put on a nice suit, button it on the front. Stand up like a soldier in front of the judge and never admit to have sold anything. You will not get punishment, you do not even need to hire a lawyer, the state gets you one. Suppose the lawyer is an idiot, does not open his mouth. We do not risk going to court counting on a lawyer! We count on ourselves. "I did not sell." Nobody can prove anything then. I have been selling this stuff for 20 years, true, but in court "I did not sell!" If you play stupid like you do not have a clue about what happened at the shop, you will convince the judge. I play it really well but the judge knows me by now. I have been caught thirteen times, even the judge sees me and says "Welcome." This is how corrupted the situation is.

Fırlama has become somehow blasé about it, and although he stresses the dangerous side of his profession from time to time, he is keen on emphasising its down-to-earth side. It might look serious to be confronted by corporations and police, but in real life bribes smoothe out everything (Fırlama rarely mentions the darker side, the organised crime and threats of death, for he does not see himself as part of this story). This is normal life, after all.

For his part, Fırlama decided that what he is doing is not a crime and he is not a criminal. He supports his own position by pointing out the inflated character of the brand:

> Isn't it a crime to sell what cost you €2 for €40? Or is it a crime to sell it for €3? You decide. I never try to take anybody's money. I do not have any bad records with banks or the tax office. I do not have a problem with the state or people in person. I do not mess with people, but with the brands. Because it is my major. Like you have your masters degree, I have this. I am in the imitation business and I only use a cell phone, no emails and such. You can give my full name to Turkcell to learn my phone number, that is how unworried I am. I don't feel the need to keep my number secret. I do not try to run away, I am not ashamed that I do imitation. If a company does not feel shame to make this product for 10 liras and sell it for 200 liras, then I am not ashamed to make the imitation of it and sell it for 4 liras. They are making fools of people. They are the ones who should be ashamed of themselves. If they do not want us to imitate they should make something that we cannot imitate, that we do not dare to imitate. Then we would admire the product. But when the product is nothing special, it is only the name of the brand that is sold.

> They are selling the name. I am selling the same name for cheaper but the same quality.

Moreover, instead of the derogatory label that lawyers, usually, use to designate goods like his, that is, counterfeited brands (*sahte markalar*), Fırlama prefers the term "imitations" (*imitasyonlar, taklit markalar*), a term that indicates at least something of the work and value going into their production. However, there are also moments in which his attitude turns from confidence and defiance to bitterness. "I heard that nowadays, if you type a name on the internet, all the lawsuits of that person will pop up. I have many, so many. One might think I am a bad person. But I do imitations."

After a big, early shift from conformity and illegality, Fırlama has found a niche for himself and, through it, a morality that he is comfortable with, and that has much smaller oscillations between conventional morality and illegality. He thus talks about honesty and reliability, the quality of his products, and the complicity of the state in the development of the gray economic area in which he operates. Like many informal/illicit entrepreneurs from other parts of the world who inhabit and work within the interstices between formal/informal and licit/illicit and who are aware of the ways in which mainstream society marginalises and criminalises them as the unemployed poor, smugglers, counterfeiters, or mafias, Fırlama strives to justify his way of earning a living to himself and others (Millar 2008; Roitman 2006; Toranzo Roca 1997). Moreover, he points out that he is an integral actor in the production and reproduction of the global economy. The way in which ordinary actors, like Fırlama, make sense of their work can help us understand the present "realities of econom*ies* and legalit*ies*" (Galemba 2008: 23) and, I would add, moralit*ies* (emphases added).

Conclusion

Making fake branded clothing is an activity like a "spark," I was told in Istanbul. Among the connotations this image might raise, one could refer to the producer himself, a positive connotation of sharpness, energy, and creativity. This article makes a similar claim: The producer of imitations of branded goods might be regarded as an economic and cultural agent.

However, Fırlama's position is a paradoxical one. He has acquired sophisticated knowledge of the material strategies that are invested in the production of clothing brands. As technological upgrading under the influence of global clothing industry became possible, he engaged in the manufacturing of quality imitations. He has made his own contribution to consolidating the idea that the brand name is the paramount ingredient of a desirable commodity but, at the same time, he has come to disregard the myths with which brands are layered. He has built his reputation as an honest and knowledgeable trader and, simultaneously, grown aware of the negative connotations and moral condemnation associated with his activities.

In a way, Fırlama's dilemma is an individual version of the societal dilemma of embracing and, at the same time, condemning imitations. Fake brands are part of a pervasive material presence whose history spans centuries, whose distribution covers vast areas, and which moves between social strata. These objects exist because "we admire the unique, then we reproduce it: faithfully, fatuously, faithlessly, fortuitously" (Schwartz 1996: 16). However, in spite of this pervasiveness, fakes are denigrated. A deeply ingrained cultural logic of "either/or" binaries privileges one term at the expense of the other. The fake is caught in such an opposition and is subordinated to the original. And faking is regarded as an inherently second-rate and potentially shady activity. However, this binary logic coexists with an understanding (and appreciation) of the shades, for producers and consumers as well as lawyers favour one position or another, depending on the circumstances and their agenda. The transactional properties of fake brands are based on material properties, logo and label included, and interpersonal relationships between manufacturers, buyers, and customers. As clothes are the kind of goods whose quality can be ascertained before purchasing, and clothing is a domain in which informal connoisseurship can be achieved, the assumed separation of brands and fakes into different regimes of value becomes problematic.

The work of this producer of fake clothing brands also reveals the contradictory nature of brand. Fake branded garments are easy to produce. Sometimes all it takes is stitching a label on a piece of garment. Fake branded garments are easy to produce because standardised clothes are easy to manufacture. Furthermore, fake brands are easy to produce because of the transfer of technology and knowledge that takes place in a manufacturing site of the global clothing industry.

The endless proliferation of brands is not only a matter of available technology and know-how but also of ceaseless recycling of remains of the brand production. Therefore, it is the brand itself that creates its "other." For the brand promises quality and distinction, and, at sites of its manufacture, the inevitable concomitants are fakes and failures. The former is kept at a distance with the help of international property rights legislation. The latter is set aside through the definition of quality. Though manufactured with different intentions, fake to pass for brand, failure to be brand, in the end, the two are similar, materialising the other. Brand then constructs itself against this other, differentiating itself legally and technically.

Fake brands circulate within an uncontrollable alternative economy of value whose character demands deeper analysis, as the upgrading of manufacturing capacity in the global clothing industry, growing informal connoisseurship, and changing relationships with materiality challenges the usual assumptions and require a more diverse and encompassing perspective (Miller 2008: 23) on branding's contradictions and compromises.

Note

1. Acknowledgements: The field research on which this chapter is based was supported by grants of the Romanian Cultural Institute and Marie Curie SocAnth programme. I thank them for this support. I would also like to thank the organisers of the Cultures of Commodity Branding conference, David Wengrow and Andrew Bevan, for their valuable comments.

References

Blacher, P. S. 1996. Les "shop-touristy" de Tsargrad ou les nouveaux russophones d'Istanbul. *Turcica* 28:11–52.

Bourdieu, P. 1977. Outline of a theory of practice. Cambridge: Cambridge University Press.

Bourdieu, P. 1984. *Distinction: A Social Critique of the Judgement of Taste.* Cambridge, MA: Harvard University Press.

Entwistle, J. 2006. Encounters in the global fashion business. *Current Sociology* 54:704–813.

Galemba, R. 2008. Informal and illicit entrepreneurs: Fighting for a place in the neoliberal economic order. *Anthropology of Work Review* 22:19–25.

Kandiyoti, D. 1994. The paradoxes of masculinity. Some thoughts on segregated societies. In A. Cornwall and N. Lindisfarne (Eds.), *Dislocating Masculinity: Comparative Ethnographies*: 197–213. London: Routledge.

Keyder, Ç. 1999. A tale of two neighborhoods. In Ç. Keyder (Ed.), *Istanbul: Between the Global and the Local*: 173–86. Lanham, MD: Rowman & Littlefield.

Konstantinov, Y. 1996. Patterns of reinterpretation: Trade-tourism in the Balkans (Bulgaria) as a picaresque metaphorical enactment of post-totalitarianism. *American Ethnologist* 23:762–82.

Marcus, G. E. 1997. The uses of complicity in the changing mise-en-scène of anthropological fieldwork. *Representations* 59:85–108.

Millar, K. 2008. Making trash into treasure: Struggles for autonomy on a Brazilian garbage dump. *Anthropology of Work Review* 22:25–34.

Miller, D. 2008. Comment to D. Wengrow, "Prehistories of Commodity Branding." *Current Anthropology* 49:23.

Negus, K. 2002. The work of cultural intermediaries and the enduring distance between production and consumption. *Cultural Studies* 16:501–15.

Neyzı, L. 2001. Object or subject? The paradox of "youth" in Turkey. *International Journal of Middle East Studies* 33:411–32.

Pak, S-Y. 2004. Articulating the boundary between secularism and Islamism: The Imam-Hatip schools of Turkey. *Anthropology and Education Quarterly* 55:324–44.

Peters, E. D., C. R. Duran, and M. J. Pıore. 2002. Learning and the limits of foreign partners as teachers. In G. Gereffi, D. Spencer, and J. Bair (Eds.), *Free Trade and Uneven Development: The North American Apparel Industry after NAFTA*: 224–45. Philadelphia: Temple University Press.

Roıtman, J. 2006. The ethics of illegality in the Chad Basin. In J. Comaroff and J. Comaroff (Eds.), *Law and Disorder in the Postcolony*: 247–72. Chicago: The University of Chicago Press.

Schmıtz, H. and P. Knorrınga. 2000. Learning from global buyers. *Journal of Development Studies* 37:177–205.

Schwartz, H. 1996. *The Culture of the Copy. Striking Likeness, Unreasonable Facsimiles*. New York: Zone Books.

Tokatlı , N. 2003. Globalization and the changing clothing industry in Turkey. *Environment and Planning A* 35:1877–94.

Tokatlı , N. and Ö. Kızılgün. 2004. Upgrading in the global clothing ındustry: Mavi jeans and the transformation of a Turkish firm from full-package to brand-name manufacturing and retailing. *Economic Geography* 80:221–40.

Toranzo Roca, C. F. 1997. Informal and ıllicit economies and the role of narcotrafficking. In M. B. Leons and H. Sanabria (Eds.), *Coca, Cocaine, and the Bolivian Reality*: 195–210. Albany: State University of New York Press.

Wılk, R. 2006. Bottled water. The pure commodity in the age of branding. *Journal of Consumer Culture* 6:303–25.

World Trade Report (WTO). 2006. http://www.wto.org/english/res_e/booksp_e/anrep_e/world_trade_report06_e.pdf (accessed 31 March, 2010).

Yükseker, D. 2004. Trust and gender in a transnational market: The public culture of Laleli, Istanbul. *Public Culture* 16:47–65.

Chapter 4

The Attribution of Authenticity to "Real" and "Fake" Branded Commodities in Brazil and China

Rosana Pinheiro-Machado[1]

Introduction

Statistics frequently show that Brazil and China are the greatest consumers of fake and pirated goods in the world.[2] The reason given for this kind of consumption is usually that both nations are developing countries marked by a a level of social stratification that sets poor and rich firmly apart. From this perspective, poverty, or rather "lack of money," is the main reason why subjects opt for a cheaper fake product instead of paying the higher price for genuine brands. In this chapter, I question this economy-driven logic.

On the one hand, I agree that social inequality and poverty are important factors behind the presence of fake goods in both the producing and consuming sectors. This market in fakes is crucial for the generation of jobs and income in less-privileged areas of society. It involves millions of formal and informal workers through its processes of production and distribution. For the consumption of luxury goods in particular, the acquisition of branded commodities performs an important role in mediating the disputes and interactions that occur between different classes and social groups. Brands—as portable signs that move together with the individuals that possess them—symbolise "social well being," and are associated with luxury, power, and wealth.

On the other hand, this simplistic explanation that lack of money is always behind the choice of copies over genuine goods does not explain, for instance, why some people of low income in Brazil save almost everything they make to buy real products or why some individuals of the new Chinese elite use fake goods. Furthermore, if individuals were exclusively following the pragmatic reasons suggested by this perspective on poverty, then, when shopping for a purse, they would simply choose the cheapest product. Rather, the explanation for and meaning behind the consumption of any commodity can only be apprehended when it is observed in particular social contexts, where a brand derives value and legitimacy from the negotiations that emerge within a relational universe. Given that people are choosing between genuine and fake goods, the motivations for this discerning consumption must be related to the social construction of authenticity. Thus, although in both countries the display of luxury brands can constitute a strategy for individual empowerment—whether through replicas or genuine goods—its meaning can vary substantially from one country to another, as these meanings are related to very different local and historical contexts.

The analysis that follows traces a comparison between two circuits of consumers of branded fake goods: southern China and southern Brazil. I have performed research in both countries as part of a broader project, from 1999 to 2007, which aimed at understanding the global commodity chain that starts in the factories in Guangdong Province and ends in the informal street markets of the city of Porto Alegre, capital of the state of Rio Grande do Sul in Brazil. As I performed fieldwork in such commodity spaces, with particular attention to the system of production and distribution that connected them, my attention also started to focus on the consumers, "refusing to treat production as privileged moment or phase in a story of a commodity" (Foster 2006: 289).

I will argue that between the city of Shenzhen (mainland China) and Hong Kong, brands constitute an objectification of the everyday imposition of the border and of the identity negotiations occuring between the people of Hong Kong and the mainland Chinese since reunification. In contrast, in the case of the city of Porto Alegre, Brazil, this difference in consumption practice will be relocated to the issue of class, and I will argue that branded goods constitute tangible signs of patterns of relationships and stratification between different social layers. In both of these observed universes, what is at stake is the social construction of the authenticity of portable goods and its variation according to the context.

Authenticity

To Walter Benjamin, original works of art carry an intrinsic aura of which their reproductions, even the most perfect, are deprived:

> the *hic et nunc* of art, the unit of its presence in the very place where it happens to be. ... The *hic et nunc* of the original constitutes what is called authenticity. ... What characterizes the authenticity of a thing is everything that it contains and that is originally transmissible, from its material duration to its power of historical testimony ... that which withers in the age of mechanical reproduction is the aura of the work of art. (Benjamin 1980: 7–8)

I argue that the aura to which Benjamin refers is a perception: a reaction that the object causes in individuals. Therefore, aura is not an essential element to the real piece, but a social and individual construction. In this sense, not only real pieces can have aura—imitations can, too, since aura is a belief.

The aura, thus, can be understood as a sort of magic—in the sense defined by Bourdieu (1980)—its efficacy, acknowledgement, and value are sustained in the production of belief, which is formulated in fields of disputes and in fields that legitimise authority. If authenticity is a constructed "social alchemy," it is also social consensus (Eco 1984), based on the collective belief that an object carries certain specific, unique, and rare material and immaterial properties.

Authenticity, imitation, and power have been broadly analysed by anthropologists from many perspectives. These topics are discussed both in studies seeking an understanding of cultures, societies, and individuals (Bhabha 1994; Brown 1998; Lindholm 2002; Taussig 1993) and in more specific fields, such as art/relics (Price 1989; Schefold 2002; Wengrow 2008), commodities, and brands (Friedman 2005; Notar 2006; Vann 2006). A religious relic, work of art, or commodity has different kinds of authority, defined in singular fields of dispute, able to distinguish what is *real* and what is *false* (religious groups, connoisseurs, market, state).

In this chapter, I discuss the formation of authenticity in branded commodities. In the capitalist global market, authentic brands are distinctive symbols attached to commodities that possess intellectual property rights (IPR). The owners of the brand have social legitimacy that is sustained by market and political principles.

Through the TRIPs (Trade-Related Aspects of Intellectual Property Rights) signed in 1994 by the World Trade Organization, companies acquired the right to go to court in a specific country to fight falsifications and to call on the state police for assistance. In this context, imitations are no longer considered "the sincerest form of flattery," nor do they assume the "prestigious" pedagogic character defined by Marcel Mauss (1960). They amass negative value, classified as criminal productions. As in the field of art, brand owners have the authority to reproduce authentic "first samples." "Intellectual Property Rights laws determine which copies are authorized, legitimate, and authentic, and which copies are unauthorized, illegitimate, and inauthentic—and therefore illegal" (Commbe quoted in Vann 2006: 286). The social recognition of this legitimacy, however, is not automatic.

The meaning assigned to notions such as original, real, authentic, fake, false, replica, or mimicry can vary substantially from field to field and from society to society (see Notar [2006] about China and Vann [2006] about Vietnam). I am adopting here the pair fake/real to refer primarily to the origin of the objects: whether a product (1) comes from an authorised factory or (2) is an "outlaw" copy. That does not mean that I recognise the IPR as an absolute truth that grants purity to goods, only that I am establishing a starting point to show, later, how individuals in China and Brazil interpret such an origin. In reality, following Appadurai's (1999) thoughts, I want to show how commodities in motion have floating regimes of value and how they get and lose authenticity as they circulate in the world.

In a previous study based on ethnography (2008), I dealt extensively with the fake/real relationship from the point of view of production (in Chinese factories) and its distribution to Brazilian informal markets. My objective was to show some limitations arising from the concept of imitation as well as of from the notions of legality and illegality in these circuits. Depending on *where* and through *whom* it circulated (the appearance of the market, the payment of taxes or not, the condition of the salesman, etc.), the commodity's status could metamorphosise somewhere along the chain. In the same way, in discussing the authenticity of brands, I questioned some ambiguous goods that have the ability to dislocate the polarity of real or fake. Such is the case of the "Cucci" watches, produced in small factories in China. Can they be considered an imitation? There is no consensus. The Gucci Company may accuse the factories of intellectual property infringement, but that may not be accepted by a country that recognises Cucci as a proper trademark, and therefore as a legal form of commerce (Pinheiro-Machado 2005, 2008).

With regard to the pair real/fake and from the point of view of commodity brands, it is important to consider the material properties of the copies, which vary enormously from commodity to commodity, and from factory to factory, but range from rough copies to perfect replicas. During my fieldwork, for example, I visited factories of fake watches that used the same components and parts as the product they imitated (a not very famous brand), even the same supplier. Another factory was outsourced for the production of toys and produced a certain number of authentic products on demand from a specific brand. In the same place, the owner accepted orders for copies of the same toys, but produced them with similar but cheaper materials. Likewise, copies of certain labelled clothes (t-shirts, underwear) can be very similar to the genuine articles, as their production is simple and can be done at home, demanding only some good fabric, the model of the cutting, and the print for the brand. In these cases, where the material properties vary very little between copy and original, authenticity is sustained only by power and by the political and economic authority defined in a field of disputes that legitimises the owner of IPR.

In the native speech of traders and producers of fake goods, there is a generalised and standardised discourse, which can be found in different parts of the world, based on the common idea that copies are practically identical to the real products and the only thing that varies is the price. Therefore, those who commercialise these products are benefitting society by bringing the "same" products to the market for a price that takes them out of restrictive elitist circuits and makes them popular. According to this explanation, traders and producers respond to themselves and to society, rescuing a moral dimension to what they do, which is in opposition to the idea that their work is a kind of piracy or crime (Pinheiro-Machado 2005). As people often say in Brazil, to invert the common advertising logic that "piracy is a crime," "the real crime is selling a CD for 40 reais."[3]

After some years observing the discourse of traders who work with copies and have a positive perception of them, I started to conduct research on young people in Hong Kong—who consider themselves politically correct consumers—and I faced a completely different discourse: an actual apology for the perceived lack of authenticity and quality of the copies. The hundreds of label stores around Hong Kong left me in no doubt about the appeal that such brands exert on the population. The border city of Shenzhen has some of the most famous malls in the world for replicas of branded products. The proximity of these two universes makes it clear that the study of copies cannot treat real products too cursorily, as both get their meaning and value in oppositional or complementary relation to the other.

On the same occasion, I visited the outsourced factories of Dolce & Gabbana, Armani Jeans, and Benetton in Shenzhen and interviewed representatives from Loius Vuitton. The differences in how manufacturing was organised at these name-brand factories were very revealing. It took a long time to produce a pair of jeans, and the cost of production was very high. The equipment was extremely expensive and had limited manufacturing capacity. The production of D&G purses demanded even higher levels of supervision. All of it was carefully specified by digital prototypes from the creative office and hand-made pilot examples, which generated a pilot, hand-made model. The care given to the details perplexed me. The factory could produce around 3,000 purses a month, compared to the millions existing on the fake market. One of the partners explained: "We've had the opportunity to produce Chanel and Louis Vuitton purses. For us, it's not worth it. You can see the products that we have here, those machines are extremely expensive, very expensive. ... To produce a purse with Louis Vuitton quality, we would need even [more] superior machinery. We can't."

Copies are products that, by definition, are meant to be popularised. Genuine luxury products have limited production, so they are restricted to the elite. The replicas share the dream of belonging to that restrictive circle. My point in discussing the production of fake and real commodities is that both objects can be very similar. However, with specific regard to luxury labels, the attention to detail—expressed in time and in the cost of production—is an important distinction between the real and fake commodity. Such care becomes a symbolic value, enframed in the category of "quality." The symbolic value aggregates surplus economic value. The commodity fetishism of a Chanel purse, for example, is based on the labour practices in which the factory worker spends a long time finishing the product by hand sewing.

Aware that the copies may provide a near-perfect image of the real product, the luxury brand companies sustain their differentials through the quality discourse. A replica of a Dior watch, even the most perfect one, will never have its 12 diamonds. As a weapon, the dealers in copies have the price and the accessibility of the product. "You created a dream, we popularised it," said one of the biggest falsifiers of Cartier watches, in an interview to the press from jail. The fake product is a way to sell the commodity fetish, through a simulacrum, without an extreme aggregated value.

When real or fake goods leave the factories, the price and quality values are differently negotiated and appropriated by consumers. Away from the discourse of its "owners," a brand's power of persuasion depends on its symbolic efficacy based, as Lévi-Strauss puts it (1979), in the trinity of individual, society, and sorcerer (the object). The object has to convince, the individual has to believe, and society has to acknowledge. Thus, there are two possible ways to analyse the use of brands: from the point of view of the person who consumes them (the individual/self), or from the point of view of those who observe this person (the society).

On the one hand, it is important to understand the meaning of authenticity for those who consume the object and what that reveals about the micro- and macro-contexts in which they find themselves. When is the origin (i.e., from established shop or fake market) important? For some, the fact that a product is real is vitally important; for others, what matters is just the external appearance of product. Thus, quality, image, price, and novelty are permanently negotiated categories. On the other hand, from the perspective of society in general, the authenticity of an item, beyond its origin, depends on a subject's ability to convince. Many individuals can wear extremely valuable goods, but the universe around them may not grant them legitimacy, which means that there exists no social belief that the product is authentic. In contrast, subjects who bear sufficient social and symbolic capital can be extremely convincing and can often thereby grant authenticity to fake goods. I consider it important, then, to analyse not just the individual meaning of the brand, but also its relationship with the classifying social systems that permanently authenticate or falsify objects.

Classes and Copies

Veblen (1998), Simmel (1971), and Bourdieu (1975, 1979), through different paths and conceptions, inferred that the elite constitutes a model upon which all other classes base themselves. Behaviour and fashion are born at the top and are ramified for the rest of society through cycles of imitations. From this point of view, replicas and fake goods would always be understood as a strategy by the lower classes for claiming a new social position. Such a theory premise can be an interesting way to think about certain situations, but it does not explain the polysemy of meanings that the use of a copy can assume in the contemporary world.

Bourdieu (1979), analysing differences in behaviour among the social classes of France in the 1960s, pointed out that lower class groups would consume rough imitations (substitutes) of elite objects (sparkling wine instead of champagne, for example), and that this is a defining practice that unveils the symbolic capital of each class. Taste (good taste), then, begins to be understood as a social construction and an acquired capital, of which the lower classes are deprived. To Bourdieu, the *habitus*—a set of social and cultural dispositions embodied in subjects—is able to reveal strongly marked class frontiers. In Bourdieu's model, a person from the low-income class could never pass as a member of the elite. Despite his best efforts, his posture, taste, and preferences will denounce him. Besides that, imitations would be of visibly bad quality and therefore would not pass as real items. Following this logic, a person from the dominant class simply would not consume an imitation.[4]

Juxtaposing Bourdieu's argument with the observed reality in China and Brazil, some elements are suitable for thinking about the branded objects discussed here, others are not. Nowadays, as much in China as in Brazil, consumption of branded fake goods spreads across many different social classes. In contrast to Bourdieu's assertion, fake goods are not restricted to consumption by popular groups to emulate the elites. Obviously, these differences in part reflect the fact that the market for copies that existed in 1960s (the period of Bourdieu's research) cannot be compared to the contemporary phenomenon of simultaneous banalisation and specialisation of the replica industry. In addition, developing countries such as Brazil and China have singular historical class formations that are not necessarily similar to French society.

Why, then, are fake goods today part of a consumption reality for many classes and social groups? How can we define borders between groups given the banalisation of what is supposed to be, by definition, a restricted (luxury)? Certainly imitations are not widely consumed only because they are well made but also because the differences between the social classes are strongly marked by another conditioning factor that supports the brand: the *habitus*. At this point, a Bourdian analysis is particularly elucidative, as it shows us that there are other border definers to work alongside the objects. The *habitus* is in the body *hexis*, in language, in information, in knowing, in acting. To Friedman (2005: 332), "Distinction is not simply show but is genuine 'cargo,' which always comes from the outside, a source of well-being and fertility and sign of power. … A Congolese can identify everyone's social rank crowd by their outward appearance."

Goods constitute bridges that bring together or set apart groups (Douglas and Isherwood 1979). The social legitimacy of brands will always be restrictive and excluding, pointing to games of social differentiation and ways of maintaining hierarchies. Clothing is a privileged field to perceive such differences because it is an immediate and portable expression of the self and "is not a symbol of social position but a concrete manifestation of such position" (Friedman 2005: 237).

Mainland China and Hong Kong: Separated by Louis Vuitton

What does the long line that forms around the Louis Vuitton and Gucci stores in Hong Kong tell us about the island's relationship to the mainland? How can a consideration of brands reveal latent aspects of the reunification of the two geographical parts? Can an analysis of the consumption of brands reveal latent social and cultural aspects of the reunification process?

Hong Kong was returned to China in 1997 after a century as a British colony. The reunification, which was part of the Chinese policy of "one country, two systems" (*yi ge guo, liang ge zhi*), is still a ongoing process of negotiation, part of the everyday life of the mainland Chinese and of Hong Kong people, especially among those who live in the border region that puts them in permanent contact. If, on the one hand, face-to-face relations have become much closer, on the other hand, symbolic frontiers have also emerged that separate, limit, and denounce who is who.

The British presence is still very strong and cannot be found only in the "look right/left" signs written on the urban pavement or in the double-decker buses. It is mostly present in the mentality of Hong Kong people, who have transformed this heritage into symbolic capital with which to distinguish themselves from the mainland Chinese. The fact that they have been a British colony allows those in Hong Kong to feel that they are bearers of knowledge with regard to dressing, eating, acting, and other instances of a "cosmopolitan and globalised" world. It is a construction of identity that oscillates between China and England.

Johnny, a 23-year-old music student, told me he considered the Chinese rude and ignorant with respect to many aspects of what he considered as "modernity" (such as knowledge about brands, global cities, and the cultural industry). Feifei (32 years old), my interpreter and key-informant, has lived in Hong Kong for 15 years but kept many contacts with relatives and friends in Shenzhen. During some of the meals we had on the mainland, she would always point at someone and say: "The Chinese have no background, they don't know how to eat. Their heart is purer than the Hong Kong people's, but they don't know etiquette." Chinese table manners are extremely ritualised and marked by a complex etiquette, so when Feifei refers to "lack of politeness," she evidently is talking about a Western behaviour pattern. Anna, a 17-year-old art student, used to say that although her parents were Chinese, she would never date a Chinese person, because "they are dirty, antiquated, and don't know how to treat a woman." Rude and impolite are recurring descriptive categories that came up in interviews or conversations about the next-door neighbours.

Returning to the question posed at the beginning of this section, luxury brands constitute a mechanism that materialises this difference, commoditising identities, since they refer to the imaginary notion of an international elite. One does not only consume a purse, but also diverse representations of cosmopolitanism and travel that portray Hong Kong's wealth and prosperity. It is not enough to possess a luxury item; it is also necessary to dominate other codes of behaviour, such as eating, travelling, and body care. Thus, social distinction comes through a broader social apparatus that supports the use of the real product. For the informants, it was not only about wearing the brand, but understanding what it meant and being able to distinguish a real from a fake item.

It is important to point out that the passion for brands is not an exclusive Hong Kong phenomenon; it is a noticeable general feature in east Asia, especially in Japan (see Keet 2007). In the blog "I am fashion," author "Barneys Girl," 21 years old, comments on the book *The Cult of Luxury Brands*:

> In Hong Kong, luxury consumption has been adopted as the central ideology. Making money and blowing it on the luxuries of life is the one thing that the people are clear about, while on everything else, from politics, culture to patriotism, ambiguity reigns. Coming from Hong Kong, I must say, I totally agree.

As a counterpoint to this ambiguous identity shared between China and the the Western world, the informants are attracted to the "incontestable" authenticity of the goods they consume. For them, it is important to go to the shops on the day a new model is released, wait in line until the shops are opened, feel the environment, talk to the salespeople. There is no glamour about being in those overcrowded stores, where the attendants do not have to be nice, because the number of visitors overwhelms the intended capacity of the store. Going to the shop is, most of all, the assurance that an authentic product has been bought; it is the certainty of the origin; it is knowing that the product is part of so-called legal and ethical consumption. There are cases similar to these pointed out by Foster (2007: 708) when he discusses "lovemarks": "They signal an emotional connection and attachment to a brand that goes beyond reason." For consumers who are in love with a brand, only the real goods possess the aura that Walter Benjamin (1980) described. A perfect replica is not good enough, for it will not bring with it the distinctive experience or sensation.

Although this consumption can refer to an imagined global and cosmopolitan world, it only makes sense when related to a particular universe concerning the local identity disputes. If luxury brands are well-known Asian phenomena, in Hong Kong their meaning is directly related to the next-door universe, China. The idea of "politically correct" consumption is valued because of a presumption that, very close by, there is a "politically incorrect" consumption. For example, when I inquired about luxury consumption in Hong Kong, without mentioning the mainland, I usually got an additional response that referred to the Chinese/Cantonese and their habit of consuming and producing imitations, copies, and falsifications. In Johnny's words:

> Chinese imitate everything we make. They imitate what we consume, what we wear. If we like Gucci, Gucci will be falsified in Shenzhen. They copy even the name of our square. They have their Times Square, which is an imitation of our Times Square.

Ian, a 26-year-old young man, asserted:

> I spend lots of money buying real brands because I find hateful the consumption of fake goods. I would never buy one. It's a matter of civilization. You know, rules, capitalism, law. ... Fair-trade. But if you go to our neighbouring city, everything you are gonna find is fake.

This kind of answer, which automatically evokes the mainland, offers a way to reinforce positively their own consumption because the exaltation of authenticity needs an inauthentic opposite. To Hong Kong people, the belief that they are more polite than Chinese people is based mainly on the consumption of real goods instead of the repugnant fake ones that clearly symbolise a "less-civilized" universe. In this sense, brands are today one of most important ways of commoditising differences. To wear Gucci is, above all else, a way in which people understand and make themselves Hong Kong people.

Leaving the subway and crossing the Hong Kong/Shenzhen border, you see Lou Hu Commercial City, one of the biggest malls in China and one of the biggest centres for brand falsification in the world. It is a gigantic, modern construction. The city entrance for those coming from

Hong Kong announces a completely different universe, at least with respect to the value assigned to brand authenticity.

Shenzhen is an extremely new city, which, in less than 30 years, has turned from a village to a metropolis. During the period of economic reform, and as a result of its location on the border with the former colony, it was made a Special Economic Zone of the country, with the objective of bringing the mainland and Hong Kong closer together. The economic reforms by Deng Xiaoping that began at the end of the 1970s led to phenomenal development in Chinese industry. Accelerated industrialisation brought with it booming production of fakes, as this constituted one of the easiest and fastest ways to generate income and work. The copies are produced in the industrial zone along the delta of the Pearl River, Shenzhen's region. The population of this city is considered the "Chinese window to the world" and is composed of "floating populations." These include migrants from all over the country who make up the industrial labour force, state employees, and a large number of businessmen, who, along with the technocrats, comprise the country's new financial elites (Li 2000).

For almost a decade, China produced these goods without a strong internal market. However, internal consumption became inevitable due to its astounding trajectory of economic growth (Davis 2000). Therefore, a series of official campaigns were launched in the 1990s that encouraged people to start consuming. It was necessary to cultivate the pleasure of individual consumption through the idea of "modern and sophisticated consumers." Thus, the consuming desire was generalised through a top-down initiative. This consumption revolution in China, or the "conspicuous movement," started as a state political strategy to connect the national economy to the global economy (Pun 2003). According to Davis (2000), from 1978 to 1990 per capita income doubled, and luxury goods, which once were rare and acquired through relationships with state officers, became available to the masses.

In the last few years, the consumption of luxury brands (real or fake) in China has become an important phenomenon. From my point of view, brands are icons of a rupture with the sober Maoist aesthetics of the Cultural Revolution and introduce the country into a new phase marked by a global capitalist market economy. They strongly represent the post-Mao China, announcing a new period of wealth and welfare. When people use such products, it seems that they are constantly communicating the fact that "I live in a new country," and, indeed, this is what they actually say.

In Shenzhen, as a matter of fact, it was possible to find people wearing fakes throughout the territory, but these could be in the hands of a North American businessman, a Hong Kong person, any Shenzhen inhabitant, a factory worker, or even a homeless person or a beggar—which is extremely common. Inside this polysemy of meanings regarding the importance of brands, I have met people who bought only real products, only replicas, or both, depending on the situation. There was no homogeneous discourse about brands, such as I found in Hong Kong, but diluted expressions of understanding brand authenticity.

Whereas Hong Kong people used to tell me that the Chinese were "less polite," the mainlanders, in turn, classified their neighbours as conceited and carrying a sense of superiority, though this was not common. According to Feifei's cousin Mei-xei, 26 years old: "They think they are superior just because they belonged to England. The problem is that they will have to carry forever the burden of being Chinese. They are just like us, but think they are not."

Mei-xei takes a lot of care of her appearance and, as many Chinese women do, dresses and wears make-up in the style of the fashion models who circulate around Tokyo, Hong Kong, Beijing, and Shenzhen. Because her salary was low (1,400 yuan, or about $200 USC), she used to go shopping in the many malls that sold not only imitations but also clothes and products of cheap and unknown brands. She once told me that if she had a boyfriend from Hong Kong it would be possible for her to get presents from the established shops. However, soon after that she said that this was not so important, as she already had a "Miu-Miu purse" and a "Louis Vuitton watch." The products she referred to, nevertheless, had been bought in a famous fake market. The pride she felt on showing her belongings to other people—"Look, I have a Louis Vuitton watch"—made me suspect that, for her, the distinction between the fake and the real product was malleable. She was sharing a legitimate symbol and that was what mattered, beyond the intrinsic properties—that could be equally valuable but never decisive.

Having said this, she made lots of effort to acquire real Gucci products from her boyfriend. These were like trophies to her, because she was having an affair with a guy from Hong Kong. She used to say: "You will see! He will give me one!" The importance of genuineness in that situation resided in what it meant to the boyfriend, as the guy had never bought anything fake for himself, but had for her. Metaphorically, it was as if the purse's certificate of originality was also a certificate for the relationship: The purse was an *authentic* measure of affection. When she realised that he would not buy it for her, she said (in front of an established Gucci shop, looking at a new model in the display, thoroughly disappointed): "I don't need any guy to pay anything to me. This is an old behaviour of traditional culture. I can buy by myself. It's not the real one, but it is Gucci as well." And then she bought one at the fake market.

In another instance, in a meeting with successful businessmen that I participated in through my research on production in China, I noticed that all men were wearing a Rolex watch—one of the symbols of belonging to the new Chinese business elite. That brand, like Gucci in Hong Kong, is a form of display that empowers people to control anxiety in face of a new life circumstances. After that meeting, one of the entrepreneurs, Li (36 years old) took me to a huge market of fake watches and said: "Look my Rolex there!" The tone in which he talked about his watch was similar to Mei-xei's. The fact is, he had a Rolex. From the point of view of many Hong Kong people, he probably did not have one, because the aura of the brand for them was only attached to the genuine piece. For most of my informants from Shenzhen, however, the aura was the symbol itself, legitimised through social circuits and in moral values and beliefs.

Mei-xei and Li's cases have less to do with their salary and more to do with the general role of brands in post-Mao China, where national concepts about copies and intellectual property are very flexible. Fake goods are less criminalised than in other places; the moral weight around them is far lighter. This condition is related to the very notion of intellectual property in a country that, according to Alford (quoted in Vann 2006), does not share the North American and European traditions of recognising rights of ownership. In this sense, for many subjects, it is possible to have both real and fake goods depending on individual negotiations and values because, for this type of consumers, brands do not exist just for the real product but also for the fake one. Thus, replicas can be authentic branded commodities.

Brazil: "It's Not Because I'm Poor that I Don't Like Good Stuff"

A series of classic Brazilian sociological studies, which drafted a general theory about the country, have shown that, behind the ideas of the "cordial man," "harmony among classes," and "racial democracy" (which produce an image of a friendly country and one without strong social hierarchies) lies hidden a society marked by social stratification and racial segregation. Intense contact between unequal segments of society is only possible once subjects have a clear view about the place they occupy in society and in this interaction. It is possible to "be informal" not because there are no distances between individuals but because these distances are well known and incorporated (Buarque de Holanda 1997; DaMatta 1981). My objective here is to consider the use of luxury brands in this Brazilian sociocultural model, where society is marked not only by class stratification but for the most part by personified representations of this difference: It is important to have power and to show it.

The consumption of fake luxury goods can be found in all layers of Brazilian society. This is particularly so because it is a society that operates with a cruel discriminatory system that is able to make the accusation of "fake" for a product without any material justification. In other words, independent of whether a product is real or fake, its social legitimacy and authentication will be determined, in the end, by *who* is wearing it. This situation is not unique to Brazil, but in Brazilian society, this classification system through appearance is unusually rigid.

To exemplify this idea, I report a commonplace scene that takes place in middle- to high-class Brazilian neighbourhoods. The boss and the maid get into the same elevator of a building, both holding a Louis Vuitton purse. Not knowing the individual histories of these women, it would at first seem sensible to conclude that the first has a real product and the second a fake one. This prosaic elevator scene is an extremely useful model with which to think about the social construction of a product's authenticity. An outsider would probably assume the maid's purse is not authentic. However, there is a reasonable chance that her purse is a real item, and there are many different places where she could have bought it or received it as a present. In fact, there is a better chance that the boss is wearing the fake item, not because such branded goods are beyond the budget of middle-class citizens, but because a fake product is more likely to be successfully passed off as a real item if worn by a person with high social and symbolic capital.

My ethnography in Porto Alegre was performed among street vendors, traders who work in the informal economy. They sell cheap Chinese products as well as fakes (clothes, house appliances, decoration, CDs and DVDs, toys, etc.), and the consuming public can be from any segment of society, depending on the product. The traders belong to the low-income class, but, though they endure frequent financial hardships, are not in extreme poverty. Most of them live in the city *favelas* (slums) or in outlying cities. After some years researching the process of selling cheap products and/or fakes, I started to pay attention to what they, as traders, wished to consume.

These traders are used to selling cheap low-quality products and being portrayed by the local media as "vandals," responsible for the spread of pirated goods in town. This condition has led them to feel repelled by the goods that they sell and instills in them a corresponding desire for the expensive real product, which has to be purchased in many installments, involving a lot of effort and sacrifice. The importance of these products that they call the "good stuff" (i.e., products bought in shops) is something generated in the very community where they live in and in the place where they work.

My informants had bought DVD players long before many individuals who belong to middle class had done so (including me). They always have the most modern cell phones available on the market, a possession that is not cheap in Brazil. Amongst many examples of such consumer behaviour, I recall one in particular that addresses Miller's assertions (1998, 2008) about the intersection between consumption, relationships, sacrifice, and love. Once, a street vendor gave a very expensive designer suit to his father-in-law. Knowing the habits of that family very well, I knew that the suit would hardly ever be worn, but this did not make the gift any less important or less valuable. The son-in-law had bought the expensive and unusual gift because it had been offered to him by someone who had stolen it from a shop display. He paid half in cash and half in commodities. Without knowing his father-in-law's size, he bought the suit immediately because he knew that the most important thing was that he was giving an expensive, branded gift. The father-in-law was also satisfied with this grandiose act of sacrifice and love, although he was aware that he would never fit inside that strange and large suit.

Appreciation for brand products or new technology is a well-observed fact in other lower class contexts in Brazil. In the city of Porto Alegre, Scalco (2008) has noticed that young people from the city's biggest *favela* take part in a virtual community called "Adidas Is Style," and they prefer the Puma sports brand, displaying their taste for branded gear. Scalco also pointed out that young people would rather wear a fake Louis Vuitton than an unknown brand because, as the informants say, "on the picture nobody can realize that the purse is fake!" (p. 45). Leitão (2004) had shown that women from the same community do not like using the clothes they produce—which are bought by the middle class as politically correct fair trade consumption—preferring to go shopping in the city's big clothing stores.

The market in stolen products is one possible way to access valuable goods and luxury brands, especially in places such as the street vendor's market and the *favelas*. However, the greatest majority of people manage to buy some goods of this kind because of a very characteristic feature of the Brazilian market that is adapted to the population's purchasing power: payment in many installments. It is not unusual for someone planning to acquire a certain product to set up a savings account or to give up consuming something else, as daily food, for example. What for a long time was considered a demonstration of "ignorance" by the lower income classes in Brazil—who were accused of choosing to buy a television over food—is today seen from another perspective, in the sense of understanding that consumption is an important means by which individuals feel like socially included citizens.

My key informant, Carminha, a 39-year-old street vendor who sells toys, once spent a whole month of food money to give a branded rucksack (backpack) to her teenage daughter. Unfortunately, it was not enough money and she could not continue to pay all the installments, so she had lots of problems in court. Not shy about it, she told me what she had said to her lawyer: "Poor Catia, how could she go to school with that old rucksack?" She explained to me that at the school where her daughter studies, she suffers discrimination at the hands of other kids, who publicly call her a "smuggler's daughter." She told me: "So Catia arrived with the new rucksack and the envious kids said it was fake. ... Then I got pissed off. I went there and I almost beat the shit out of the mother whose girl said that. It's not fake, it's good stuff!"

In a society that discriminates over social class, the consumption of expensive goods, genuine and/or purchased in formal shops, appears as a mechanism for social affirmation but mostly as a kind of morality. Thus, consumption is at the centre of the process of interaction between the different social segments in Brazil. In this sense, following Friedman's remarks (2005) about the use of French brands by poor segments of society in Congo, I do not believe that such consumption of valuable goods is a simplistic process of elite emulation, but "a consumption of highest orders of status [...] it is not representative but constitutive of social identity" (p. 238).

For those working in the informal economy, or somehow related to these practices, other forms of insertion in an established universe are sought after, forms that drive them away from the labels of illegality and "dirt" that surrounds them. Evidently, these people also consume fake goods such as CDs, DVDs, games, etc., but there remains a positive narrative surrounding real branded goods that provide social affirmation in the face of other social groups. If it is true that not many grant legitimacy to a Louis Vuitton purse worn by a maid or sold by a street vendor, it is exactly in this context of dispute that buying real goods make more sense. From the individual point of view, branded goods are appreciated as they signal the possibility of having "good stuff," or, as it is commonly phrased, "Poor people also like good stuff."

For Brazilian informants, in contrast to those from Hong Kong, for example, it is not important to go to the store and live through the whole shopping experience. The means by which the goods were acquired makes no difference: hence, the market for stolen products, second-hand goods, or those paid in installments. The important thing is to own, enjoy, and show the goods off in a specific social setting. The informants wished to somehow share in a universe of wealth from which they are typically excluded, and the feeling of inclusion is achieved through the commodities.

At the market where I developed this ethnography, I perceived that many people from the middle class asked the street vendors where they could find the "Louis Vuitton purses"—a brand with great appeal among the Brazilian elites. In reality, I understand that the people who get the most out of the replica market in Brazil (and elsewhere) are certainly from the middle class—those who often do not have sufficient economic capital to buy a real purse but nonethless possess sufficient symbolic capital to legitimatise a replica.

These segments of society need to convince society that their replicas are real. For this goal to be successful, the individuals have to master certain fundamental codes of behaviour that support the purse (voice tone, body *hexis*, clothing). They are conscious that a lot of effort and even dissimulation is needed to make this act convincing. One of Carminha's old customers told me: "Ah, I went to Paris with very little money, on a tourist package, only eating sandwiches! I say I bought my Louis Vuitton there, but I actually bought it from the Koreans." Whenever I say I study "fake goods," people often respond in a low voice, with a little bit of shame, as if revealing a secret: "Mine is fake."

According to a replica supplier from Porto Alegre who belongs to a once wealthy family:

> We, purse specialists, can distinguish a replica from a real item. But I can use both and nobody can tell the difference. If I go to a fancy party with a replica, nobody is going to say that "I" am wearing a replica, there will probably be many replicas and real items mixed together, but that doesn't make any difference, what matters is that all of us are wearing Louis Vuitton and that's what matters at these places. And the people who buy from me, for instance, is an A class public. People with a lot of money.

In these spaces, where brands circulate and appear at parties and prestigious events, it is not possible to say that everybody believes in the authenticity of everyone else's goods. Purses and watches are legitimised all the time. The eyes carefully focus on the jewelry, the clothes, the personal and family background of the individual, and then a verdict is pronounced. But there is always a certain doubt in the air, an uncertainty: Is it real or fake?

For those who do not actually belong to the elite, such goods are an attempt to negotiate the class condition. So, a Louis Vuitton is an attempt to get an *upgrade,* to try to "look rich." A maid or a street vendor could hardly achieve that. In a society where class inequalities are highly marked and where power is strongly attached to the person, branded commodities, whether real or fake, constitute not only fashion accessories, but also accessories for social positioning and identity. Luxury brands, in Brazil, are in the centre of a (symbolic) class struggle.

Conclusion

In this chapter, I have presented case studies that pertain to the branded commoditisation of identities, classes, boundaries, hierarchies, moralities, and prejudices. The plot was composed of micro-stories, almost snapshots, with the objective of questioning the construction of authenticity and the value of branded goods in completely different contexts.

I showed that among certain groups of Hong Kong youngsters, from both middle and high classes, paying a lot of money for a product bought in a store depended on an understanding that, by doing so, one was acquiring certain properties that the copies did not possess—a type of aura that came along with a certificate of authenticity from an established shop. The "latest model" is anxiously awaited, and the old ones are thrown away. The importance of genuineness can be also seen among certain lower class groups in Brazil, too, where it becomes a moral value. Quality is understood as being "good stuff" and not "junk stuff." In contrast to the priorities of the informants from Hong Kong, the means by which these Brazilians acquire their goods do not matter: They could be from the stolen goods market, second-hand, or paid for in long installments. They had to be "real" products, even if they were old or *démodé*.

Such representations show that different groups opt for paying a lot of money for a product for cultural reasons, which are defined in a relational universe that deals with some Other from whom it is necessary to distinguish oneself. In Shenzhen, for example, I have found more flexible discourses and diluted values over the importance of brands, which could be attached to the copies or to the real products. In the same way, among some individuals of the Brazilian middle class, I realised that the origin of the brand was not taken into account. The important thing was the simulacrum itself, and, in this sense, the fake item was a good strategy to acquire branded goods. "Look at my Louis Vuitton/Rolex … ," some informants told me, pointing at their replicas. In these situations, the question that remains is: Is the brand attached only to the genuine piece?

Charles Lindholm, discussing authenticity and modernity, completes his argument by suggesting that "the sacred is where you find it" (2002: 337). The anxiety to define what is or is not authentic encourages truths to be created and recreated incessantly within particular fields of dispute. The authenticity of a branded commodity is linked to international hegemonic politics that classifies them as legal or illegal according to criteria of IPR and of registered brands. However, in the field of the human experience with objects, such definitions are re-elaborated, taken on or not. To answer the question above—regarding the brand and the genuine piece—I take the liberty of "pirating" Lindholm's sentence and suggest: The brand is where you believe it to be.

Notes

1. Acknowledgements: This project had the support of the Wenner Gren Foundation and of Brazil's Ministry of Education (CAPES Foundation). I thank the organisers of the Cultures of Commodity Branding conference, David Wengrow and Andrew Bevan, and the valuable comments

of Daniel Miller and participants at that conference. I am grateful to Miller's students for discussing a draft of the chapter. Ed Ross and Clarissa Rahmeier also helped with suggestions and criticism. I thank Professor Ruben George Oliven for all his encouragement.

2. According to recent reports from International Intellectual Property Alliance (2008), Brazil and China are the greatest consumers of pirated softwear in their continents.
3. Forty reais are approximately equivalent to $25 USC. A falsified CD copy costs an average of 5 reais ($3 USC).
4. A limitation of Bourdieu's (1979) analysis is that he does not reflect on the powers of manipulation and creativity of the lower classes. To Burke (1989), Bourdieu does not consider two-way movements (e.g., where the popular becomes erudite); according to Ginzburg (1987), the concept of "lowering" of goods, material and immaterial—as well as the idea of imitation between classes—is too mechanical, for it does not take transformation into account. When this appears in Bourdieu's work, it is always with the meaning of "distortion."

References

Appadurai, A. 1999. Introduction: Commodities and the politics of value. In A. Appadurai (Ed.), *The Social Life of Things, Commodities in Cultural Perspective*: 3–63. Cambridge: Cambridge University Press.

Benjamin, W. 1980. *A Obra de arte na época de suas técnicas de reprodução*. São Paulo: Abril Cultural.

Bhabha, H. 1994. *The Location of Culture*. London: Routledge.

Bourdieu, P. (with Y. Delsaut). 1975. Le couturier et sa griffe: Contribution à une théorie de la magie. *Actes de la Recherche en Sciences Sociales* 1:7–36.

Bourdieu, P. 1979. *La Distinctio n: Critique Sociale du Jugement*. Paris: Minuit.

Bourdieu, P. 1980. The production of belief: Contribution to an economy of symbolic goods. *Media, Culture and Society* 5:261–93.

Brown, M. F. 1998. Can culture be copyrighted? *Current Anthropology* 39:193–222.

Buarque de Holanda, S. 1997. *Raízes do Brasil*. São Paulo: Companhia das Letras.

Burke, P. 1989. *Cultura Popular na Idade Moderna*. São Paulo: Companhia das Letras.

DaMatta, R. 1981. *Carnavais, Malandros e Heróis*. Rio de Janeiro, Zahar.

Davis, D. 2000. Introduction. In D. Davis (Ed.), *The Consumer Revolution in Urban China*. London: University of California Press.

Douglas, M. and B. Isherwood. 1979. *The World of Goods*. London: Routledge.

Eco, H. 1984. *Viagem na Irracionalidade Cotidiana*. Rio de Janeiro: Nova Fronteira.

Foster, R. 2006. Tracking globalization: Commodities and value in motion. In C. Tilley et al. (Eds.), *The Sage Handbook of Material Culture*: 285–302. London: Sage.

Foster, R. 2007. The work of the new economy: Consumers, brands, and value creation. *Cultural Anthropology* 5:707–31.

Friedman, J. 2005. Globalization and localization. In J. India and R. Rosaldo (Eds.), *The Anthropology of Globalization*: 233–46. Malden, MA: Blackwell Publishers.

Ginzburg, C. 1987. *O Queijo e os Vermes*. São Paulo: Companhia das Letras.

Handler, R. 1986. Authenticity. *Anthropology Today* 2:2–4.

International Intellectual Property Alliance. 2008. *Estimated Trade Losses due to Copyright Piracy*. http://www.iipa.com/ (accessed 23 June, 2009).

Keet, P. 2007. *The Tokyo Look Book*. Tokyo: Kodansha.

Leitão, D. 2004. Roupa pronta é roupa boa: Reflexão sobre gosto e hábitos de consumo de produtoras e consumidoras de uma cooperativa de costuras. Paper presented at the XXIV Reunião Brasileira de Antropologia (24th Brazilian meeting of anthropology, Olinda, 12–15 June).

Lévi-Strauss, C. 1979. *Structural Anthropology*. Middlesex, UK: Penguin Books.

Li, C. 2000. "Credentialism" versus "entrepreneurism." In K. B. Chan (Ed.), *Chinese Business Networks. State, Economy and Culture*: 86–111. Singapore: Prentice Hall.

Lindholm, C. 2002. Authenticity, anthropology, and the sacred. *Anthropological Quarterly* 75:331–38.

Mauss, M. 1960. *Sociologie et Anthropologie*. Paris: Presses Universitaires.

Miller, D. 1998. *A Theory of Shopping*. Ithaca, NY: Cornell University Press.

Miller, D. 2008. *The Comfort of Things*. Cambridge: Polity.

Notar, B. E. 2006. Authenticity anxiety and counterfeit confidence: Outsourcing souvenirs, changing money, and narrating value in reform-era China. *Modern China* 32:64–93.

Pinheiro-Machado, R. 2005. La garantía soy yo. Redes locais e transnacionais de comércio popular—camelôs e sacoleiros (Brasil-Paraguai). Ph.D. dissertation, Federal University of Rio Grande do Sul, Porto Alegre, Brazil.

Pinheiro-Machado, R. 2008. China-Paraguai-Brasil: Uma rota para pensar a economia informal. *Brazilian Review of Social Science* 67:45–76.

Price, S. 1989. *Primitive Art in Civilized Places*. Chicago: University of Chicago Press.

Pun, N. 2003. Subsumption or consumption? The phantom of consumer revolution in "globalizing" China. *Cultural Anthropology* 18:469–92.

Scalco, L. 2008. Falakenois: Etnografia de um projeto de inclusão digital entre jovens de classes populares em Porto Alegre. Ph.D. dissertation, Federal University of Rio Grande do Sul, Porto Alegre, Brazil.

Schefold, R. 2002. Stylistic canon, imitation and faking. *Anthropology Today* 18:10–14.

Simmel, G. 1971. *On Individuality and Social Forms*. Chicago: University of Chicago Press.

Taussig, M. 1993. *Mimesis and Alterity: A Particular History of the Senses*. London: Routledge.

Vann, E. 2006. Limits of authenticity in Vietnamese consumer markets. *American Anthropologist* 108:286–96.

Veblen, T. 1998. *The Theory of the Leisure Class*. New York: Prometheus Books.

Wengrow, D. 2008. Prehistories of commodity branding. *Current Anthropology* 49:7–34.

Chapter 5

The *Real One*

Western Brands and Competing Notions of Authenticity in Socialist Hungary

Ferenc Hammer

Introduction

> [W]hilst the boundary divides and frames, the margin blurs distinctions and frontiers. It is one of the paradoxes of dress to be both a margin and the boundary at one and the same time. As boundary, it frames the body and separates from the rest of the social world, thus functioning as a kind of container or wrapper. Dress as a boundary is meant to trace a neat line between self and other: the limitation of physical visibility via clothing, for example, parallels metaphorically an intended limitation of psychological accessibility. As margin, on the other hand, dress connects the individual to other bodies, it links the biological entity to the social ensemble and the private to the public.
>
> —Warwick and Cavallaro 1998: 17

In this chapter, I offer a cultural-historical notion of brands in socialist Hungary, particularly with regard to how various spiritual, psychological, material, social, and political considerations all supported a seemingly unlikely phenomenon: A small niche of brand culture in a Soviet bloc–planned economy appeared a decade or two before the fall of state socialism in 1989. The core question of my research[1] addresses what jeans may have signified in the social and cultural context of a communist economy characterised by the obvious absence of marketing (advertising), regular shortages of goods, and indeed the sheer absence of Western commodities from official retailers. The main empirical source of my research on jeans-wearing in socialist Hungary is composed of 140+ written testimonies from respondents to my inquiry about the story behind their first pair of jeans. Amongst this extremely interesting material, I have noticed a recurring comment to the effect that, when obtaining their first jeans, it had been really important for many people that these should be a real brand (often Levi's and that this idea of the "real" (i.e., properly branded pants) was connected to various forms of authenticity and salience in the respondents' personal life during the 1960s–80s. These rich accounts of first jeans reveal an intriguing construct of authenticity in modernity.

Background: Dress and Politics

The "nationalization of the everyday" played an integral role in the power exercised by the newly established Stalinist regimes of eastern Europe after World War II, and this even involved control of everyday manners and habits. In *The Captive Mind*, Czeslaw Milosz (1990 [1953]: 54) analyses the everyday practices of deceit and camouflage people used as means of survival:

> It is hard to define the type of relationship that prevails between people in the East otherwise than as acting, with the exception that one does not perform on a theater stage but in the street, office, factory, meeting hall, or even the room one lives in. Such acting is a highly developed craft that places a premium upon mental alertness. Before it leaves the lips, every word must be evaluated as to its consequences. A smile that appears at the wrong moment, a glance that is not all it should be can occasion dangerous suspicions and accusations. Even one's gestures, tone of voice, or preference for certain kinds of neckties are interpreted as signs of one's political tendencies.

Milosz's eloquent account reveals two important aspects for my topic. First, he captures a unique feature of life in a dictatorship; that is, political submission, or resistance, for that matter, can be expressed by nonverbal acts of communication, such as bodily gestures. The nontheatrical performances described by Milosz are crucial aspects of political culture because these politically relevant, everyday acts are the very moments through which institutional power either reproduces itself or fails to do so. Second, he suggests that—rather surprisingly in the modern era—a modern state pays attention to its citizens' neckties. The Hungarian socialist regime's (1948–89) control over its citizens' lives can be exemplified rather accurately via changes in sartorial regulations enacted from the 1950s to the 1980s. State propaganda and the opinions of the school and workplace all fiercely attacked the rock "n" roll outfit in the early 1950s, and the ever-present accusation that this clothed the agent of the enemy could quite often turn into a legal accusation (and indeed, conflicts that arose around men's palm-tree patterned yellow ties attest to Milosz's observation). However, by the 1980s, dressing, with the occasional exception of the punk look, had become a private matter.

Blue jeans as a consciously worn fashion item started to appear in larger cities from the early 1960s,[2] and as was true in many other countries, wearing denim became, together with rock "n" roll or beat music, an integral part of a booming youth culture centred around acts of performance, immediacy, spontaneity, and a certain "lived authenticity" vis-à-vis "manipulative" and "violent" adult universes offered by capitalism or the Vietnam War. Alongside many similarities—such as school and family conflicts about proper dressing—jeans-wearing differed in communist countries in crucial ways from that of the West. Jeans as an obviously Western outfit played several suspicious roles in the political plots envisioned by various pockets of officialdom during the 1960s: Let these clothes be agents of bourgeois decadence, market-driven moral decay, and selfish aesthetic indulgence. After about 1970, jeans started to appear in the officially controlled visual spaces (public events, theatres, schools, record covers, films) and, somewhat later, at university exams and schools ceremonies or on the television screen. Since the state-owned retail industry, until as late as the late 1970s, did not regard its role as the fulfillment of teenage denim dreams, the norms regarding denim were guided—instead of ads, commercials, or shop windows—exclusively by interpersonal experience, in the form of gossip, hearsay, personal stories, or the stirring image of a passerby.

Denim-Order Basics

The collectively maintained aesthetic and normative order regarding jeans-wearing was developed independently not only of the state, but more stunningly, also of the market and local media. This truly folkloric modern consumption movement was organised around certain strictly observed rules, all associated with particular physical features of the blue jeans: branded or brandless, fake or original, position in the brand hierarchy (Levi's, Lee, Wrangler, Roy Rogers, etc.), denim or jersey, real-looking but "soiled" by a dubious brand. Or whether it is new or of a worn look. Or denim versus corduroy. Or the colour, the buttons, the pockets, or the cut. This seemingly rudimentary classification system worked in reality in a much more sophisticated way. One respondent recalls some interlocking features of the denim order in those days:

> For me the ranking was: Levi's, Wrangler, and Lee, and there wasn't any other brand for me. One's choice had easily recognisable tiny details, and it wasn't just the sheer quality, fabric or cut of the pant.

> One's social background was shown by the jeans, it didn't take even a split second to work out where a person was coming from and what they were wanting to belong to. But it was also telling whether the jeans were worn-out or dirty. Or whether they had special DIY decoration with embroidery, labels, or spots. That was the start of the worn-out style. Those with worn out jeans, or where the leg was sweeping along the ground, sent a message to the world that they were free. The upper-middle-class children, like me, could achieve only that my jeans' legs weren't ironed to a crease. (B.B. 1954)[3]

This short "my first jeans" excerpt neatly captures the central point of this chapter: Jeans offered multiple opportunities for young people in the 1960s–70s to locate themselves on various cultural maps, as well as "enabling environments," magic garments that acted according to an envisioned authentic life.

Real: Individualised, Singular

A first set of psychological and social features that seem to have been associated with the authenticity of a pair of jeans are all connected to their singular, personalised physical character. For example, the denim pants sold a generation ago required a certain series of physical interventions at home, such as a generous washing to make the jeans shrink to their "regular shape" and to fix the indigo die more permanently (and thereby avoid staining the wearer's body), or some retailoring to fit the wearer. This latter was especially necessary at flea markets,[4] where there was a limited choice of sizes or when the wearers were not the ones who purchased the jeans because the purchase occurred in a foreign country.

There is a well-known Levi's advertisement from 1985–86 that depicts an individual lying in the bath in the act of preshrinking his pair of jeans (Bartle Bogle Hegarty, quoted by Nixon 1997: 293). The performance shown in this advertisement has particular relevance in a Hungarian context and is captured by the Hungarian word "beavatás" that means both an initiation, in the sense of a ritual or rite-of-passage, and and also the act of preshrinking, in the textile industry sense. The practice depicted in this image was mentioned half dozen times in the "my first jeans" interviews, as respondents recounted how they jumped into the bathtub to spend a crucial phase together with their jeans in which the latter would be shaped to the figure of the owner.[5] It is difficult to overlook the intriguing ritual allusion, particularly to baptism, in this rather functional bathroom practice. In the bathtub ceremony, both of the unifying partners go through a transformation: The pants get the shape of the owners, who, in turn, start a new period of their life marked by a pair of jeans. The act of making a piece of clothing more worn looking, to render it usable, is most obviously connected to the figure of the dandy, particularly Beau Brummell, who had his new clothes worn by his valet or his gloves wetted first to gain the form of the hand impeccably (Barthes 2005 [1962]: 67). I will return to this the instructive parallel in a subsequent section, but it is worth noting here that in both cases the outfit unifies deeply ingrained aesthetic values (a natural look) and moral values (honesty/originality) to the degree that, in both cases, the practice of taking care of clothes is the performance of authenticity itself.

The personalisation of jeans included further practices in socialist Hungary, such as decorating them with beads, embroidery, or ink inscriptions of rock lyrics or rock stars' names.[6] But the most intimate personalisation of jeans was undoubtedly the many-year process by which they were into a second skin, their form unmistakenly marking the wearer's figure.[7]

The Body and Jeans

> It is a sine qua non about them that they should be white. ... The natural place for one's feet when one is in a comfortable sitting posture is on the top of one another, or at any rate rubbing up against one another. With white spats this won't do. You must think of each foot separately. Your right foot must know all the time what your left foot is doing; it must treat it as if it were a complete stranger, as if it were someone else's foot. White spats make you foot-conscious.
>
> —About White Spats (*Punch*, McDowell 1998: 24)

Perhaps the most important theme in the history of outfits is the connection between social status and clothing. Apart from taxonomies of fabric, colour, or cut or whether or not to dress similarly, an interesting if perhaps less immediately visible topic for historians is how a particular style of dressing regulates bodily posture and movement and how this then contributes to the reproduction or subversion of power relations. Although modern Western clothing is mostly about how to present the young(ish) fit and attractive body (that itself is the litmus test of the character), for previous epochs body gestures had offered meanings central rather to the vision of the respectable citizen.

For Thackeray, hat-wearing was a clear sign of the position of the person in question on the social map: "If you want to understand the individual, look at him in the day-time; see him walking with his hat on. There is a great deal in the build and wearing of hats—a great more than at first meets the eye" (*The Book of Snobs*, excerpted in McDowell 1998: 106). Balogh (1984: 3) notes that men's bodily postures and movement in public changed when newer generations in the interwar period decided to drop the urban hat and the high shirt collar from their everyday wardrobe, allowing them to step out from a cautious and rigid regime that made them look like carefully moving statues characterised by "the slow glance of the nobility" (in Bourdieu's words, 1989 [1979]: 177). The humorous *Punch* quote from the 1920s on why people wore white spats would make any social scientist envious as the author captures the point with exceptional clarity: White spats facilitate foot-consciousness.

Jeans create particular relationships between the body and a person's sense of self. Perhaps the most arcane but maybe the most interesting relationship appears in the "my first jeans" stories when speakers explain how their "body consciousness," together, of course, with their whole social persona, changed the moment they put on their jeans.[8] Cultural history suggests many equivalent instances in which an individual is transformed through a performative ritual that involves the donning of a particular garment. The moment following Joe Buck's purchase of his western boots in *Midnight Cowboy* suggests something similar (McDowell 1998: 97–98):

> In his new boots, Joe Buck was six-foot-one and life was different. As he walked out that store in Houston something snapped in the bottom half of him: A kind of power he never knew was there had been released in his pelvis and he was able to feel the world through it. Brand-new muscles came into play in his buttocks and in his legs, and he was aware of a totally new attitude toward the sidewalk. The world was down there, and he was up there, on top of it, and the space between him and it was now commanded by a beautiful strange animal, himself, Joe Buck. He was strong. He was exultant. He was ready. "I'm ready," he said to himself, and he wondered what he meant by that.

A good number of the "my first jeans" respondents reflect in some way on how things changed after they obtained and first put on the jeans. In these accounts, beyond the expected recollections about parent or teacher reactions to the new look, I found particularly interesting those who claimed that the act of putting on the jeans had instantly placed them in a new and different social and/or cultural position. Three respondents also mention the act of sitting on the ground in a public place while wearing their jeans. This act seemed to be more than just a question of taking advantage of durable dress: Rather, sitting on the ground sent a deliberate message that the individual was different from the older generation and from the mainsteam of the society. As one respondent put it: "People who wore jeans had an understanding that jeans provide the freedom to sit wherever one wishes, lying to the ground, in the dirt, was natural with jeans. They did not need washing" (N.L. 1958).

As this comment suggests, the act of putting on jeans opened up a new cultural and social landscape of potential roles and actions for the young people in question,[9] and this cultural space, sartorially constructed in part, was organised around values and principles such as spontaneity, informality, "lived life," and difference. This programme of 1950s–60s youth culture in Western countries and its followers elsewhere is unique in the sense that the central point of the social change is the visible representation itself: outfit, speech and lifestyle, behaviour, dancing, body gestures, etc. The consequence of this deeply "make-it-happen," or performative character of agency is that elements of youth culture were performed, as well as understood by peers, as the grand allegory of authentic life. Through a web of connotations, often connected to the publicly displayed body, blue jeans were central elements in this allegory. Just as it is quite difficult to verbalise the experience of a performative act, be it a ballet or a hammer throw, so, it is too, for a nonverbal experience of acting with the publicly displayed, carefully stylised body. Just as a close description of the athletes' outfit would tell little about their performance, the significance of jeans-wearing can be equally vaguely captured by the actual description of the dress.

The performative, matter-of-fact acting embodied by jeans-wearing both in 1950s United States and 1960s–70s Hungary can be compared to the ritual aspects of the mimetic experience in ancient drama, both enacting the deepest reality: the authentic self. In the Hungarian environment, the "mimetic struggle" embodied by jeans-wearing was not only defined in relation to a life threatened by hypocrisy, manipulation, alienation (and further popular vices of the 1960s), but jeans as a garment with a clear American origin were sometimes the target of political criticism in the official press, following the track, albeit in a milder manner, of the 1950s communist purges against the jampec subculture.[10]

The performative nature of the look of the youth had not been confined to abstract meanings about the good life, but as the "sex and drugs and rock'n'roll" slogan captured it with remarkable accuracy, the acting body had been perceived to be in the centre of the vision of making a good life. It is quite ironic that, in the West, youth culture often saw bourgeois morality and capitalist consumption as the enemy, whereas in eastern Europe this ideology of the late 1960s manifests itself in a more fragmented and faded manner and, amusingly, does so in reverse, in the form of classic Marxist product fetishism: Records, magazines, and denim pants (the latter possibly with a red tab on them) were the things used to fight the alienation and oppression of modern life.

Jeans and the Sensual Body

> But one thing is certain. When I walked down the street in my new jeans, my buddies, but particularly the girls were gazing at me like a nun stares at the Christ sculpture.
>
> —Sz.S. 1943

Jeans worn in an extremely tight-fit manner, by both men and women, represented another important element in the transformative continuum of the body, the outfit, and the social world. Again, a curious connection can be detected through the analysis of the "my first jeans" stories between the publicly displayed body, jeans, and a landslide change in sexual habits. As far as the basic grammar of this visual regime is concerned, a respondent put it this way:

> No question, jeans were all about sex. They are designed to show off the body, especially the female body, and this was especially true in the pre hip-hop era when they were worn supertight. Their cut boldly accentuates the male and, even more so, the female genitalia. In Spanish bullfighting there is an expression: three-handkerchief matador. Many people wearing jeans back in our day were three handkerchief hippies, if you know what I mean. (Sz.A.1963)

"You know what comes between me and my Calvin's? Nothing," murmured Brooke Shields in a deeply suggestive, squatting pose in a 1981 Calvin Klein jeans commercial (Craik 1994: 195). It seems that in a half generation or so, the jeaned female body had become such a visual commonplace that selling jeans now demanded the mobilisation of the (almost) invisible, "underneath" for its purposes. In Hungary in the 1960s, this would have all been unnecessary, because a girl wearing tight trousers of a male cut (with a fly!) was a sufficiently outrageous concept and look in itself: As many of the jeans stories revealed, parents (particularly fathers) and schools often joined forces against the idea of females wearing jeans until the early 1970s. Again, a great deal of fantasy and imagination played a key role for everyone concerned. A woman born in 1967 wrote:

> [Jeans suggested about the wearer] a certain level of maturity and a serious past (look at my jeans and find out why they are so worn). Associated with maturity, jeans suggested a certain amount of sexual experience, and considering the sex appeal lent by the pair of jeans, this suggestion could have been easily materialized, suggested the jeans. It was all in contrast to the trinity of skirt-blouse-high heels that was worn only by miserable, childish good little girls. (K.A. 1967)

Another woman the same age reflected on the attitude of her teachers:

> In our school, the teachers believed that girls who wore jeans did so because they wanted boys to touch them, but the teachers' propaganda became counterproductive, because eventually more and more girls found it appealing to belong to this imaginary, demonized group. (Sz.Zs. 1967)

Since the 1968 youth movement's agenda with respect to sexual politics was about liberation from patriarchal sexual oppression, jeans often provided the catalyst or the way for a woman to act freely. Jeans meant two things, at least for girls: emancipation and freedom—put one female "first jeans" respondent bluntly (K.J. 1952). Some of the jeans stories offer a glimpse at what these abstract concepts meant in reality. For example, the structuring role of dress in social relations is eloquently exemplified by this amusing love triangle story, with a female "first jeans" respondent at the centre:

> We decided with my boyfriend ... that we'd wear the same style of jeans and shirt. ... His aunt came back from America with two pairs of Lee, for two of us. ... This L. was my boyfriend at home in the small town. My college boyfriend in the city (without knowing about it) joined our jeans community. I got a shirt from him (that I already had at home), not knowing that there are already three of us wearing the shirts. (K.B. 1950)

As revealed by the following recollection, jeans were a decisive factor in the selection of company from the opposite sex: "Odd as it may sound now, [in college] I didn't even look at boys who didn't wear jeans. If he had a denim jacket also, it had even increased his chances at me" (Sz.K. 1962).

In these stories, the visible jeaned body (lower part) played a dual role. Jeans as a frame allowed the wearer to express sensual meanings with the help of the clothed body; meanwhile, the publicly displayed "properly" framed (i.e., jeaned) body had created opportunities for the subjectivity to perform changes and to refine itself. The clothed body as the mould for the subjectivity is also the very communication interface for the young self to connect with others in managing anxieties and seeking pleasures.

Jeans as Anti-Worlds

> Girls in Szeged wore different styles of jeans in the late 1970s, but all were flared. I didn't find these simple enough. I was rather austere in my norms and habits and jeans were just appropriate for its

> expression. Quite paradoxically, the simplest design was the most difficult to find.
>
> —V.V. 1959

In many ways, jeans represented an anti-world, the negation of some kind of societal norm often established by the majority in the same age group, by adult society, or even by the eastern bloc. As discussed above, the most obvious frame for understanding jeans is the absence of scarcity. Americans jeans were the ultimate outfit in the United States of the 1970s and 1980s , but, in contrast, the jeans of eastern Europeans in the same period were predominantly thought of as something *not* eastern European (but Turkish, for instance). The main imprint of the Cold War on the constructed meaning of jeans was (especially in the 1960s) that they were politically "juicy" clothing that could represent silent, nonverbal, but striking forms of political opinion. Another use of images and meanings from the "other world" is considered by Berger et al. (1974: 144):

> To take two important examples, the pair of the wristwatch and the ballpoint pen in the Third World today cannot be understood simply in functional, utilitarian terms. These objects are above all symbolic. They symbolize the modern status of the individual who owns and exhibits them, and they are well chosen indeed. The one refers to literacy and the immense new stocks of knowledge that literacy opens up, the other, to that structure of time which ... lies at the very roots of the technological production and modern bureaucracy, and thus of modern society as such.

Wearing jeans was a performance of dissent, alienation, or detachment that could be directed against not only explicit political entities, but virtually anything. A respondent (B.R. 1959) describes her preference for jeans in the 1970s that were a classic cut of Lee, Levi's, Wrangler (or a good quality copy), but definitely not one of the contemporary Italian-style baggish cuts (see chart). Similar oppositions can be discovered in her remembered choice of other consumer items as well:

B.R.	Classic cut jeans	T-Shirt, jumper, male shirt, flat heels	Long hair, let down to the shoulders	No jewelry	Arts, theatre, jazz, avantgarde	Classical music: Bartók, Penderecki
"Others"	Baggish cut jeans	Blouse, skirt, tailor-made dress, high heels	Pony tail, hairdresser-made hair	Gold jewelry	TV, light pop music, disco	Classical music: Romantics

This rough chart that I have distilled from the written narrative of B.R. shows the systemic character of her choice regarding jeans. The obvious parallel of values in the various consumption fields suggests a homological pattern between the cultural repertoire of B.R. vis-à-vis the repertoire of the mainstream youth in the turn of the 1970s–80s. In fact, the relationships between these cultural fields suggest that an "achievement" in one field (e.g., obtaining a pair of Levi's 501s) perhaps led to at least partial achievement in other fields as well, or at least the promise of it. According to this logic, putting on worn-out jeans in socialist Hungary during the 1960s–70s instantly suggested an entire programme. The elements of this programme were often rejections of the prevailing sociocultural norms of the country. It is important to see that the anti-world aura of jeans was not just the mechanical negation of certain norms, such as low heels versus high heels, but a conflict of point of view as well. While the adult world was shocked by the habit of middle-class youngsters neglecting to wash their jeans, the jeans-wearers' spectacular choice of norm regarding cleanliness was not just a sheer dirt cult, but a strategic act highlighting what was, in their view, an improper norm system that favoured good manners over the "spontaneous expressivity" of the individual. In a way, as these homologies suggest, the dirty jeans themselves indicated a sense of purity and authenticity in the otherwise vulgar, dishonest, and defaced world of middle-class culture.

The multi-layered normative nature of jeans-wearing quite naturally explains—in spite of the series of excitements, pleasures, and adventures—a degree trauma present in the "my first jeans" stories, such as conflicts, doubts, angst, and despair. Some of these negative experiences can perhaps be explained by the insatiable nature of modern needs or a sense of loss—the feeling of losing one's future—in the actualisation of an expectation (Campbell 1997). Some of the negative experiences stem from conflicts caused by the jeans in the family, at school, or within a peer group. Some troubles were caused by the mere fact that the jeans' owner was not there at the point of purchase and had no control over the choice of proper brand, cut, colour, or size. Apart from all of these funny but not too surprising factors, the sheer number of tragic features in these stories seems to underline the existential importance of jeans in the respondents' lives.

The close normative grid that connects aesthetics and morality in jeans culture, naturally invites a closer look at the figure of the dandy. As Barthes (2005 [1962]: 66–67) argues in his study of the dandy, the desire to differentiate within otherwise fairly uniform male dress of the 18th–19th centuries created the (previously less important) notion of *detail*: "which started to play a distinguishing role in fashion: the knot on the kravat, the material of the shirt, the buttons on the waistcoat, the buckle on the shoe, were from then on enough to highlight the narrowest of social differences."

In the same way that detail made early modern male dress into a complex normative interface, so, too, the surprising regularity of cultural meaning associated with jeans in 1960s–70s Hungary suggests something similar. The notion of the "half ready," "raw" jeans (before the preshrunk denim era) echoes Barthes's words about the dandy, for whom "all forms of behavior which bear witness to the profoundly creative, and no longer simply selective, idea that the *effects* of a form have to be thought through, that clothing is not simply an object to be used but is a prepared *object*" (Barthes 2005: 67, emphases in the original).

The curious orderliness—through a web of homologies—of the otherwise explicitly disorderly youth culture of 1960s Hungary shows a parallel with the sartorial system of Beau Brummell, the archetypal dandy who used a meticulous, almost absurd, selection system for his daily outfit.[11] In both cases, the organisation of the outfit performs a dual function, almost a romantic programme, in which efforts at a certain aesthetic perfection are closely connected to a moral disapproval of the ordinary and the shallow.

Real: The Change-Maker

> Perhaps it was important to obtain the jeans to prove to myself that I was willing to make changes in my life.
>
> —Sz.B. 1958

I have already discussed an interesting feature of jeans: When they are connected with the body, the "medium is the message" for the individual.[12] This motif can be understood in a wider sense, whereby jeans are not just a cultural mould (or programme) for the individuals but also an important event in the person's personal biography. The historical and sociological challenges that young people faced if they wanted to obtain jeans are the basics of Cold War history: political control on travel and consumption, shortage, isolation from Western culture and media, lack of money, a conservative society, etc. Yet, the efforts young people made to (1) evade adult oppression, or (2) obtain something valuable, in an adventurous way and from a remote place (I cannot help thinking of the Golden Fleece here) seem to be timeless conditions associated with preparing for adult life.

This dual character of the ordeal, revealed by many of the "first jeans" stories, explains the crucial role of jeans in the life of lots of teenagers in the era. Buying jeans with the proceeds from a first salary earned as a summer student worker, challenging parental or school authority via jeans-wearing, and many other conflicts all have a common element in the difficulties surrounding jeans: They were catalysts for young people to define their changing position in adult life, securing a crucial connection between the clothing and the identity of the individual, even if this retrospective narrative ("making one's way into adult life") is presumably different from the jeans-seeker teenager's narrative decades before for whom getting a pair of jeans was everything but a step toward adulthood. Though brief, the following comment from a 40-something man's story might be compared to more spectacular generational changes: It shows that washing a pair of jeans still can represent something of a threshold a good 30 years later: "That was the first garment that I washed with my hands—and I did that with pleasure, I have to add" (P.Z. 1964). The "first jeans" stories record how wearing jeans brought a long-awaited change in the male owner's status among girls,[13] or how starting at a new school was considered to be possible only in "real jeans":

> Finishing elementary school in the summer of 1989, just before the drivers' blockade, I felt that one could not start highschool without jeans. The feeling that everyone had them except me, was obvious for me, but honestly, I don't know whether I was really the only one in Kaposvár, the center of the world at that time for me, who was strolling the streets in fake jeans. (S.T. 1975)

The personal biography of these individuals is sometimes framed by the histories of their jeans. Jeaned time often contains of series a plans, adventures, and complications associated with acquiring the item, then comes life with jeans, life in denim, and also occasional mention of a jeans afterlife, when they were made into shorts, a guitar case, or just became no longer used memorabilia in the wardrobe. Another slightly extraordinary aspect of jeans that connects them to the various life cycles of individuals is that they were quite often inherited from older and/or larger cousins and friends, a habit otherwise not particularly frequent among young adults, or they were passed on in a similar manner. As is clear in many of these jeans memories, properly branded jeans and the rites of passage of a young person were naturally connected through the notions of the real and the authentic.

Just as youthfulness had started to become a trait marking a distinct social group in post-World War II modern societies—with a particular set of aesthetics, morals, and aspirations—so, too, the perceived importance of a chosen cultural style—in music, clothing, slang, haircut, etc.—had started to grow. As the jeans stories suggest, getting a pair of jeans was slightly more than just a necessary "uniform" for the membership of a desired group. The act of putting them on was remembered both as an act of transformation and as an act of stepping into a desired status within a peer group. In the "first jeans" narratives, lovers remember having denim pants made—based on their own design—by a local tailor; grey-haired men remember, not without emotion, when their friends lent them their jeans for a few weeks; and numerous further stories testify to the happiness of stepping into the "in" crowd with a pair of new jeans on. At the turn of the 1970s–80s, two parallel youth cultures—with their conflict similar to the mod-rocker relationship in the United Kingdom some 20 years earlier—flourished, each with a particular choice of denim. Although Italian-style jeans—of Casucci or Ritt brands—were worn by youth with an affinity for disco, decent haircuts, and pointed shoes, the followers of the "rugged style" expressed a preference for rock music, long hair, and less decency in behaviour and wore traditional jeans with their legs retailored to an extremely tight fit.

Authenticity and Nonlinguistic Denim Experiences

> In his scent, the whole person seems to be suddenly there; not just his words or thoughts, but the whole physical person and everything that makes him attractive, desirable and *him*. By means of smell, a whole atmosphere, person or situation comes to you in a very direct and unmediated way. And a comparable power of attraction proceeds from objects.
>
> —Veenis 1999: 79

It is a puzzling task to consider the role and significance of nonlinguistic experiences in a culture that is—no matter how important the visual has become in the past few decades—organised predominantly through the medium of spoken or written language. An understanding of the roles of nonlinguistic experiences—the most notable in this respect is the olfactory—in our cultures has to take into account a paradoxical double problem of representation. On the one hand, the status of the nonlinguistic as such (the olfactory, the tactile, our relationship with the visual or with objects) in modernity is inferior to words. For example, the nonlinguistic experience, such as the history of smells, or the role of the tactile has a quite esoteric aura as a social science topic (not least because it is largely lacking in written sources). On the other hand, when our culture takes into consideration the nonlinguistic experiences, again, for example, the olfactory, it can be approached through language only (through what else, one could ask), therefore the nonlinguistic is "too representational," because our culture finds it difficult to accept it on its own right. If they become subjects of attention, they can be grasped through only vague and ambiguous linguistic references. The result is, that, for example, our relationship (knowledge and norms) regarding smells and odours is framed by linguistic techniques (metaphor, metonymy, simile, etc.).

Another epistemological difficulty stems from the fact that the nonlinguistic has a different temporality to the experiences captured by spoken words or texts. One can recount how an event took place in the past, but when one smells or touches something strongly associated with a past event, that thing, though it would be placed in a past tense for narrative, becomes somehow present. One cannot overestimate the importance of this atemporal experience for the self and subjectivity: A flash-like glimpse of something long forgotten with the help of a less temporal nonlinguistic experience is a strong existential event.

The role of the nonlinguistic in the overall politics of memory is discussed by Connerton (1989) in terms of incorporated bodily practices that "record" history for participants differently than the way it would be recorded by historians. The "first jeans" stories suggest that people's memory of jeans, with their strong connection to bodily experience, and to the materiality, tactility, and smell of jeans offer a very different history than the ones in history book about everyday life. The less temporal and less representational connection to objects and stronger emphasis on tactile and olfactory experience can be detected in numerous cases in the denim narratives. A man recalls his first encounter with Levi's jeans:

> I still remember well when I saw real Levi's the first time, as a small town boy. An older boy from Budapest came to the neighbors, and he wore that famous garment. I was appalled by the noble material, the luminescent indigo, the copper rivets, and generally the whole look. Its cut, following the wearer's body, the strong stitches, the legs not ironed to a crease represented a revolt against everything that we were fed up with: Wearing suits, older generations, the official cheesy pop music, clumsiness. (V.Gy. 1951)

The attraction of the materiality of jeans can be found in lots of denim recollections. This feeling has obviously to do with the exoticism, the scarcity, the Western origin, the political allusions, and many other social and historical factors. But it is likely, too, that the frequent references to materiality are, in fact, the very moments when the speakers do not only remember with their words, but by nonlinguistic means re-act events with their body and psyche, too, as if they lived in a continuous present.

Materiality in the denim stories is, of course, not only important because it make possible alternative relationships with the past, but also because this materiality presented the clearest dividing line between the East and the West. Veenis (1999: 105), in her above-quoted study, describes how former GDR (German Democratic Republic) citizens expressed their attraction to Western objects that were "somehow softer, brighter, more colourful and cheerful and how their scent was better than that of the East German things." This difference between Western and Eastern products meant that a key skill of the eastern European consumer-citizen was to spot instantly the Western origin of things. This skill naturally included the ability to differentiate between jeans of Western or Eastern origin. Perhaps the picture of Hungarian-made jeans shown in Figure 5.1, with the poor quality of the stitches and the clumsy cut, can explain that this skill was far from any supernatural faculty (see XXI. század 2006).

The scarcity and price of jeans and the possibility of counterfeit items lent the issue of real versus fake a special importance in the denim stories. The anxiety and insecurity associated with unknown brands, the price of jeans, or what various cuts are supposed to look like created an amusing culture of authenticity procedures:

> Of course jeans went through a rigorous test in the school. First it was rubbed with paper tissues to see if the paper becomes blue. Then followed the cigarette test on the bottom of the pant's leg to see if it gets burnt or not. The last one was the mechanical test that was lifting the jeans owner by the belt holders of the pant. (K.T. 1950)

In this section, I have explored the significance of an extremely unusual set of connections encouraged by the materiality of the jeans, the strong physical and psychological relationship between the body and dress, and the anxieties about the proper qualities and origin of the jeans. In this

Figure 5.1. A jeans-like outfit made in Hungary in the 1970s.

intriguing human-material relationship, issues regarding the authenticity of denim allowed subjects to define themselves in relation to important questions in young people's lives at that time.

Authenticity through Possession: Having *Therefore* Being

Anyone who has ever seen 1–2-year-old children's instinctive and passionate sense of property with respect to a toy or a blanket—without ever having been taught—can not accept at face value any grand dictums that property is predominantly violence, theft, or sin, because as much as our personal history is commonly accepted as a fundamental defining structure of our self, the same could be argued for the objectified record of our life (i.e., things that we choose, live with, and preserve). In the jeans stories, again, economic factors (shortage, supply-demand imbalance, etc.) might explain all too well the respondents' common reports of their emotional outburst following the acquisition of a pair of jeans and may cast doubt on possible sociocultural motifs behind the act of possessing.

Jeans stories sometimes suggest fundamental motivating factors behind the desire to possess things, other than scarcity.[14] Odd as it may sound, a lot of the denim stories make strange allusions to an "expanded self," or "real self," after they tell of success in obtaining a pair of denim. Just as most likely toddlers' attraction to their "property" is somehow connected to their emerging body consciousness (and perhaps to an instinctive territoriality), so the celebration of the self through possession a pair of jeans has something to do with the fact that jeans have an exceptionally strong relationship with the body. This psychological understanding of property can explain the changing roles of tactility in modernity.

Touching things in a market might be fairly a preemptive effort to avoid buying rubbish, but also it can be understood as a ready substitute for possession. Through tactililty, the observer becomes participant, and because the body is utilised in the process of investigation, touching things can "really" tell the buyers what it would be like if the object was their property. This psychological and anthropological approach to objects, which emphasises their constructed character, offers a broader understanding of the relationships that exist between humans and objects and one that is different from or more than just the meaning given by passive and dumb object. Jeans stories suggest that jeans can serve as an enabling tool, a prosthesis that perfects the acting self, or—as an important memory—a crucial event that contributes to the shape of the memory of the individual. It is important to note here that the starting point for my jeans research was when I noticed that the vast majority of my friends and acquaintances between roughly 30 and 65 years old have a well-preserved story of their first pair of jeans.

It is interesting to consider how, if one accepts Marx's dictum on value to the effect that it "transforms every product of labour into a social hieroglyphic," the reading of a product made in a communist economy would differ from the reading of hieroglyphs embodied by a capitalist-made product. The following picture offers a few guidelines to assess how objects could mean slightly different things in socialism and capitalism. Figure 5.2 shows a Soviet-made kitchen garlic press, made perhaps around the 1970s. The most remarkable feature of this object is that its price (1 rubel and 80 kopeks, equal to about $3 USD officially, or .30 on the black market) is cast into metal.

Casting the price of a product into metal is a wholly utopian act, an expression of the belief that all those contingencies that may affect the price of the product and can be bracketed. There is more, however. A capitalist economy and market renders as relative not only prices but also shifts in supply and demand, brand values, and/or fashion appeal. Without the relativising function of a market economy, product features can seem so fixed and eternal that they turn products into idols and remove them from determinations of time and space. This observation is echoed by Michel Pastoureau (2001 [1991]) in his illuminating book on the history of stripes, arguing for the more "matter of fact" significance of public appearance in premodern societies compared to modern capitalist ones, where "all fixed

Figure 5.2. A garlic press made in the USSR in the 1970s.

and fast frozen relations are swept away," including the meaning of outfit: "What is true for clothing in actual society is even more true in iconographic society. Here more than anywhere else, dress is a medium for signs and a tool for classification" (p. xiv).

The idol-like features of branded products in socialism, caused by the lack of a market economy that melt all solids into the air, might partly explain the distinctive position of jeans memories in respondents' recollections as well as those recurring motifs in jeans stories that suggest that having and being are not opposites—as it is suggested by moralist arguments—but rather that the former is the means to achieve the latter.

Flügel (1930: 29), in his classic *The Psychology of Clothes*, partly referring to Spencer's observation, highlighted a salient aspect of dress is its potential connected to the concept of a trophy. Though this kind of psychologising logic is quite distant from the mainstream of my argument in this chapter, the notion of trophy nonetheless captures quite neatly some important aspects in my respondents' "first jeans" stories, such as the obvious "display" potential of the eternal, rare, and valued thing, the hardships experienced in acquiring jeans, and the perceived change of the status of the acquirers of jeans amongst their peers, all presenting the acquirer of the trophy with the status of being exceptional. But as suggested in the previous section about eastern European citizens' hypersensitivity to objects of Western origin and the ability to identify them from a great distance, one can understand brand mania not only as a preminent form of status seeking, but also as a visible demonstration of someone's rejection of the socialist value system.

Remembering and Lived Authenticity

The main source for my research is the collection of first-person memories about jeans, quite often citing events from the 1960s–70s. Remembering events at several decades distance is already a risky enterprise in terms of source accuracy, and, as a regime change took place between the act of remembering and the event in question, it is even more so. The key politico-epistemological issue about the remembrance of socialism is that those living their (young) adult life in the 1960s–80s and choosing not to immigrate illegally mostly assumed that the political system would not change in their lifetime. This led not only to massive public surprise and amazement concerning the speed and the smoothness of the political transformation, but also resulted in a massive memory problem: In personal memories as well as scholarly articles, events of the 1970s–80s are often framed as "precursors" to the subsequent fall of the Iron Curtain, whereas in reality, even in 1987 Hungary, very few thought that the Soviet system (including the Soviet military occupation) would change in the foreseeable future. Interestingly, people's recollections about their first pair of jeans often manage to elude this trap of remembering by offering no particular moral of the story. Paul Betts's (2000: 753) study offers an interesting observation on the role of objects in remembering (although branded jeans were not socialist products, their "mythical" era was confined to the period ending in the 1980s): "Despite—or perhaps precisely because of—the abrupt 'secularization' of GDR artifacts, where they no longer embody the dreams of a prosperous present and a hopeful socialist future, they now serve as repositories of private histories and sentimental reflections."

Written personal jeans memories offer few internal lines of evidence as to the factual credibility of their account or the degree to which they have been, quite unavoidably, streamlined by their narrator. What is very obvious, however, when browsing the dozens of first-person accounts is an overall tone of seriousness, not apparently due to any sense of ceremony but because the respondents recognised a stake in such reflections. Many respondents stated explicitly how unexpectedly pleasant it was both to recollect jeans memories and to realise how well they remember a story that they had not thought about for a long time. Reliving the past via means of remembering fits quite neatly with the frequent allusion in the denim stories to the fact that the wearing of jeans was often about living a life more real than before. This reflection, from a now 60-year-old college professor, suggests that the ever-evolving, mutual interrogation between jeans and the self is a never-ending story in his life:

> If I have a choice, I still wear jeans, T-shirt, jumper and sneakers. ... Jeans wearing doesn't mean for me a distinct feeling, or belonging to a group, or a feeling of youthfulness, but if I try to find out why exactly I still wear jeans, I really cannot come up with anything new other than these things I mentioned that would explain why I put on jeans again and again. (M.S. 1949)

Conclusion

Obtaining a pair of Levi's, Lee, or Wrangler was typically less a capricious fancy and more the result of a sometimes months-long complicated plot, so the question of authenticity (real brand versus no-name brand, real jeans versus fake) was connected in my respondents' stories through personal, political, aesthetic, psychological, and sociological patterns of authenticity in people's life. In this short analysis based on 140+ denim narratives, I have identified some major interlocking patterns associated with the question of authenticity and its relationship to the body, sexuality, the nonlingiuistic association with objects, the singularity of the jeans, the materiality of denim, temporal connections between the biography and the jeans, or community and belonging.

It is crucial to remember that this rich set of meanings related to authenticity and involving the human and the material should not suggest that jeans were only about authenticity. As I have pointed out (Hammer 2008), jeans were also sources of exclusion, conflict, conformity, and privilege. Jeans, both as an interface connected to the body, and an interface connecting the self with others in the social world, had served as a time- and space-setting instance that moderated change in people's life. This research offers both an opportunity to take a fresh look at the roles brands fulfill in people's life—through situating brand-related meanings in a thick network of meanings associated with the material, the psychological, the visual, the social, and remembering—and an alternative to the all-too-well-known Cold War accounts of blue jeans and communism. Remembering occasions of wearing a particular outfit, contrary to conventional forms of memory connected to largely verbal experiences, offer illuminating opportunities to reenact something in the past, and therefore, for a split second, evading the authority of the passing time.

Notes

1. Acknowledgements: I conducted the substantial part of this jeans research with the help of a research fellowship from Birkbeck College's Cultures of Consumption programme in the spring of 2006, for which I am grateful.
2. Interestingly, as some of the "my first jeans" respondents noted, blue jeans started to appear sporadically after the 1956 Hungarian revolution, in aid bales and clothing packages sent by Western relatives, but these odd work pants or costume-like trousers were regarded as so extravagant—called "cowboy pants" at the time—that wearing them in schools or outside of the home was usually out of question (see more detail on this in Hammer 2008).
3. Initials and the date of birth indicates the origin of the "my first jeans" story excerpt.
4. The main place to buy jeans in Hungary from the 1960s–80s was the Ecseri second-hand flea market on the outskirts of Budapest where customers could buy their jeans in rather romantic circumstances, for example, trying pants on next to a Biedermeier armoire. Ecseri's jeans merchants were criss-crossing the border of the legal economy: Though many of them had a merchant's license, their merchandise had to be smuggled in from abroad, by truckers, conductors on trains with sleeping cars, or other nonelite travellers.

5. In their study, Kaiser and Ketchum (2005: 139) cite case-studies that show how this formulation can in fact be reversed. Jeans shape the body as well, when they function as a measure of the owner's diet-watching efforts.
6. The significance of the individually decorated jeans can be understood as a kind of personalised uniform. A typical phenomenon in military barrack culture is when soldiers, otherwise wearing the same uniform, develop a sophisticated hierarchy through individualised uniform-wearing styles, including careful choice among several different available uniform items and individualised personal military gadgets.
7. A small but telling contribution to this point is the story of a bank robber who was identified by the authorities based on CCTV pictures through the "ridges and valleys" of darker blue along the seam on the leg of his jeans, and that the FBI had included the "fingerprint" of worn jeans in its identification database (Hauser 2004).
8. This complex relationship is often suggested in various literary traditions by the notion of the talking dress, such as it appears in this section from "Jean Rhys' Illusion" (Adams and Tate 1991: 4):

   > Wonderful moment! When the new dress would arrive and would emerge smiling and graceful from its tissue paper.
   >
   > "Wear me, give me life," it would seem to say to her, "and I will do my damnedest for you!" And first, not unskillfully, for was she not a portrait painter? Miss Bruce would put on the powder, the Rouge-Fascination, the rouge for the lips, lastly the dress—and she would gaze into the glass at a transformed self.

9. Mick Farren (1985: 8) shares a similar personal recollection in his book on the black leather jacket. Just as jeans seemed to have been a ticket to sit on the ground in public, for the author, the jacket was the entry point into the world when he would smoke in the presence of an adult:

   > As I stared into the mirror, I couldn't believe myself. Admittedly the mirror was tilted up to produce the most flattering effect, but I looked great. My legs seemed longer, my shoulders seemed broader. I flipped the collar up. I looked so damned cool. Mother of God, I was a cross between Elvis and Lord Byron. The old guy who was taking care of me asked how it felt. I lit a cigarette. I'd been smoking for maybe a year or so, but this was the first time I'd ever lit one while I had an adult's total attention. Jekyll was already becoming Hyde. I told him grandly that it was okay.

10. This was a kind of mixture of the Zoot Suiters and the Teddy Boys in the 1950s in Hungary (Hammer 2008).
11. He had his coat, waistcoat and trousers made at three different tailors, and he was famous for spending a good part of the morning tying his cravat properly (Harvey 1995: 29).
12. It is quite instructive to apply some of these formulations of the "medium is the message" thesis to clothing; see, for example, these short excerpts, all from the first paragraphs of McLuhan's *Understanding Media:*

    > [T]he personal, and social consequences of any medium—that is, of any extension of ourselves—result from the new scale that is introduced into our affairs by each extension of ourselves, or by any new technology. ... What we are considering here, however, are the psychic and social consequences of the designs or patterns as they amplify or accelerate existing processes. For the "message" of any medium or technology is the change of scale or pace or pattern that it introduces into human affairs ... , the medium is "the message" because it is the medium that shapes and controls the scale and form of human association and action. (1964: 23–24)

13. K.T. (1950) said: "As soon as I put on the jeans, at least I felt that way, I could approach girls with more ease and relaxedness, and the girls seemed to feel more like accepting my approach too."
14. Or they explain why the noun "goods" —in Hungarian as well as in English—comes from the adjective "good."

References

Adams, B. and T. Tate, Eds. 1991. *That Kind of Woman*. New York: Carroll and Graf.

Balogh, Z. 1984. Egy marxista öltözködésszociológia vázlata [An outline of a Marxist sociology of dressing]. *Medvetánc* 16:3–32.

Barthes, R. 2005 [1962]. Dandyism and fashion. In A. Stafford and M. Carter (Eds.), *The Language of Fashion*: 65–69. Oxford: Berg.

Berger, P. L., Berger, B. and Kellner, H. 1974. *Homeless Mind*. New York: Vintage Books.

Betts, P. 2000. The twilight of the idols: East German memory and material culture. *The Journal of Modern History* 72:731–65.

Bourdieu, P. 1989 [1979]. *Distinction: A Social Critique of the Judgement of Taste*. London: Routledge.

Campbell, C. 1997. The romantic ethic and the spirit of modern consumerism. Reflections on the reception of a thesis concerning the origin of the continuing desire for goods. In S. M. Pearce (Ed.), *Experiencing Material Culture in the Western World*: 36–48. London: Leicester University Press.

Connerton, P. 1989. *How Societies Remember.* Cambridge: Cambridge University Press.

Craik, J. C. 1994. *The Face of Fashion: Cultural Studies in Fashion.* London: Routledge.

Farren, M. 1985. *The Black Leather Jacket.* New York: Aberville Press Publishers.

Flügel, J. C. 1930. *The Psychology of Clothes.* London: Hogarth Press.

Hammer, F. 2008. Sartorial manoeuvres in the dusk: Blue jeans in socialist Hungary. In F. Trentmann and K. Soper (Eds.), *Citizenship and Consumption*: 51–68. Basingstoke, UK: Palgrave Macmillan.

Harvey, J. 1995. *Men in Black.* London: Reaktion Books.

Hauser, K. 2004. A garment in the dock; Or, how the FBI illuminated the prehistory of a pair of denim jeans. *Journal of Material Culture* 9:293–313.

Kaiser, S. B. and K. Ketchum. 2005. Consuming fashion as flexibility: Metaphor, cultural mood, and materiality. In S. Ratneshwar and D. Mick (Eds.), *Inside Consumption*: 122–43. London: Routledge.

McDowell, C. Ed. 1998. *The Pimlico Companion to Fashion.* London: Pimlico.

McLuhan, M. 1964. *Understanding Media.* New York: Signet Books.

Milosz, C. 1990 [1953]. *The Captive Mind.* New York: Vintage International.

Nixon, S. 1997. Exhibiting masculinity. In S. Hall (Ed.), *Representation: Cultural Representations and Signifying Practices*: 291–330. London: Sage.

Pastoureau, M. 2001 [1991]. *The Devil's Cloth: A History of Stripes and Striped Fabric* (J. Gladding, Transl.). New York: Columbia University Press.

Veenis, M. 1999. Consumption in East Germany: The seduction and betrayal of things. *Journal of Material Culture* 4:79–122.

Warwick, A. and D. Cavallaro. 1998. *Fashioning the Frame: Boundaries, Dress and the Body.* Oxford and New York: Berg.

XXI. század. 2006. Dzsínsz. RTL KLUB Television.

Chapter 6

Royal Branding and the Techniques of the Body, the Self, and Power in West Cameroon

Jean-Pierre Warnier[1]

In one strict sense, a brand is a red-hot iron. Branding is done by pressing a brand sign into the surface of a thing or onto the skin of the person to be branded. The mark that is left often cannot be removed. Branding has been applied in the past to human beings to mark them as slaves, as convicts, or for a ship's galley. It is still done to cattle. Branding is applied to a surface, to the skin. It is a practice that is related to the process of commodifying or decommodifying a person or thing. A cow may be branded so that its owner may claim ownership, and/or sell it for his or her own benefit. Branding a slave for the royal galleys removes that person from the slave market. It should come as no surprise, therefore, that, in dealing below with branding in an African kingdom, I start with the description of a mother working the skin of her toddler. This will provide a case on which I will subsequently expand.

Massaging a Child

In December 2002, I watched and filmed the massaging of a one-year-old boy by his mother. It was in the Western Highlands of Cameroon, also called "Grassfields" (for an overview, see Warnier 1985). It took place around 11 a.m., when it was becoming warm enough to give a bath to the child in the open. The mother brought a plastic tub into the courtyard of her compound. She put it on the ground next to a chair. She also brought a plastic basket full of various cosmetics and some warm water that she poured into the tub. She added some Dettol into the water as a disinfectant. Then she gave the baby a bath. After that, she sat on the chair, lifted the baby boy from the tub, dried him with a towel and laid him on his belly across her own lap. Then she took a container of Pears Baby Lotion from the plastic basket. She poured some into a smaller container the size of a drinking glass. She then took hold of a container

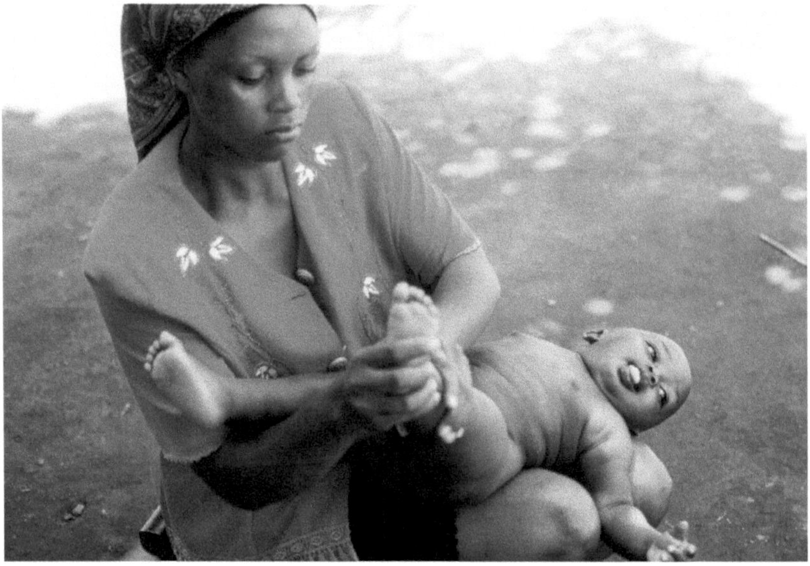

Figure 6.1. Massaging a baby (2002, author's photograph).

of Pears Baby Oil, added some of the oil into the small container, and mixed together the oil and the lotion.

Then she proceeded to massage the boy. She sat him on her lap, put some of the mixture in the palms of her hands, rubbed her hands together, and started massaging her baby, beginning with his scalp, which was shaven, and working down. She took more mixture into her hands as they got dry. She worked every square millimetre of the baby's skin, taking great care when rubbing the ear folds, the nostrils, the folds in the skin, the buttocks, the genitals, and the spaces between the fingers and the toes. As she went, she shifted the baby. She also worked and turned all his joints around (Figure 6.1). The whole business took about 15 minutes. The baby was relaxed and calm. After that, she dressed him with clean pants and a t-shirt and sat him down on the ground.

The woman had performed this type of massaging to her baby every day since his birth. She intended to do so until he would be able to walk. Most if not all women do likewise in the Grassfields. All the women use branded/labelled cosmetics for that purpose. The most popular brands at the time of my fieldwork were Pears, manufactured by Unilever in Nigeria for the African market, and Bébé Hygiène.[2] There were half a dozen more branded baby lotions, including Johnson's (Figure 6.2). Each woman had her own preference. Some, for financial reasons, used locally manufactured palm oil, which is much cheaper. In the past, it was the standard body lotion for babies and adults as well—mostly women. Palm oil was also extensively

Figure 6.2. Branded cosmetics and disinfectant in use in 2002 (author's photograph).

used for cooking. It came from different places outside of the kingdom, and people differentiated the oil according to its place of origin, its technique of manufacture, colour, taste, etc. As a body lotion, it has been almost entirely displaced by industrial ones. As a cooking oil, it is still a common staple.

The Techniques of the Body, the Skin, and the Self

I now want to relate cosmetics and palm oil to various concerns with the body and its skin. Massaging a baby is one of the techniques of the body practised in the Grassfields. Many of those bodily techniques apply to the skin as an envelope, to its openings and its contents.

The techniques include: shaving the scalp and other parts of the body, rubbing the skin with various substances, applying pigments such as camwood and kaolin, washing with decoctions, introducing substances into the skin by scarification, introducing substances through the skin's openings (mainly the eyes and the mouth) and, until the advent of colonisation, administering the sasswood ordeal to suspected witches by having them absorb a mixture of the pounded bark of the tree *Erythrophlaeum guineense*.

The substances used are: water, raphia wine, decoctions of vegetable substances "washed" with water or raphia wine, camwood (a red pigment taken from the wood of the tree *Pterocarpus soyanxii*), red palm oil, the oil extracted from palm kernels, a mixture of oil and camwood, saliva, various kinds of medicines, ashes from charred animal and vegetable substances, a mixture of saliva and medicines or raphia wine, animal fat (the most valued one being that of the python), a mixture of medicines and fat or oil, white or pink kaolin, and all kinds of cosmetics purchased on the marketplace.

Brands clearly matter in this context, but it is not easy to know in what way. It depends on sensory perceptions that are notoriously difficult to verbalise, and it also depends on origins and the way they trigger various fantasies. People are choosy. I have little doubt that advertising is important in the recent history of consumerism in the Grassfields and in other parts of Africa (see Burke 1996; Presthold 2008). However, in the Bight of Biafra and its hinterland, successful brands or advertisements for cosmetics usually capitalise on bodily techniques that are widely shared in the whole area, especially when it comes to the skin. In "Wes Kos" or pidgin English, "*ao fo you shikin*?" means "How for your skin?" (i.e., "How is your health?").

As many primatologists and psychologists such as John Bowlby (1958) and Didier Anzieu (1985) have noticed, the skin is extremely important as a surface of contact with the outside world, as a container for internal organs and as a means to exchange and vet heat and cold, UV rays, moisture, sweat, food, drink, noises, and the like. The external senses—sight, taste, smell, hearing, and touch—are located around its surface and openings. From a psychoanalytical point of view, in the ontogenesis of the subject, the earliest experiences of the foetus and the newborn with regard to its skin surface allow him or her to construct its psyche via the fantasy of an envelope with an inside, an outside, specific psychic contents, and a capacity for introjection and projection.

This is, so to speak, the universal substratum behind many cultural variations in the way human beings deal with the skin. What is peculiar to the Cameroon Grassfields is that this universal dimension of the human experience has been put to use to construct a monarchic political organisation. In the Grassfields especially, the skin is involved in power relationships and I elaborate on this point below.

There are some 150 kingdoms in the Grassfields. To this day, the king's body is perceived as containing a number of substances (saliva, breath, speech, semen). These substances are extended or multiplied by other substances (palm oil, raphia wine, camwood), exterior to the king's body but attached to it through the use of royal embodied containers and by specific bodily techniques (e.g., the king puts raphia wine into his mouth and sprays it onto the people; see Warnier [2007, 2009] for further details). These substances are perceived as being transformed into ancestral life substances when the king makes offerings to the dead monarchs while uttering performative words. The burden of kingship consists of storing, accumulating, transforming, and giving out material substances as life substances. The king acts as a container. People receive those substances on the surface of their skin as a container or within their bodies through its openings. Royal power is addressed to the body of the king and of the subjects.

The king is the one who constructs the inside, the outside, the openings, and the one who ensures the circulation of substances, things, and people through the openings. He also acts as the opening in the container. The king may be considered as having (or being) three bodies—his own personal body, the palace, and the city—as these three bodies have a similar material structure: an envelope with its openings, irrigated throughout by the same royal substances. The king achieves the closure and opening of the three spaces by so many envelopes, their openings, the difference between an inside and an outside, the circulation of substances, things, and people through the openings, taking in the "good" things and expelling the "bad" ones as so many excreta.

The government of the kingdom may be a matter of making decisions and implementing them. But, above all, it is a matter of opening up, closing down, pouring in or out, and taking in and pushing out material things and people in transit, that is, commodities. As regards taking things and people into the kingdom and storing them, or pushing them out, the king is in charge of branding everything that is in transit through the openings. How does he achieve this? He applies royal substances and words on their surfaces or skins. He does this himself or has it done by palace notables and stewards on his behalf, using royal substances such as camwood powder and palm oil from the royal stores. Spraying his own saliva or a mixture of saliva, breath, and raphia wine on things and people may be seen as the epitome of his action in branding things and people.

In Foucauldian parlance, we are facing a governmentality of containers. The techniques of the self apply to the skin, its contents, its openings, and to the way things go into it and out of it again. They belong with the technologies of power.

Looking Back into the Past: Standardized, Packaged, and Labelled Commodities

Why is it necessary that the king imprint his mark on things and people in transit through the openings of his three bodies? I would argue that, in the past, the Grassfields were characterised by high population densities, a monarchic political organisation, the practice of long-distance trade between notables and kings, and a regional trade in subsistence goods for several centuries (Rowlands 1979; Warnier 1985: 9–173). This political and economic organisation has carried on well into the 20th century when industrial consumer goods began to be added on top of the local commodities or actually replaced them. In addition to practicing intensive trading, people were, and still are, highly mobile, and kingdoms were and still are composite organisations federating descent groups and kinship groups that do not share any common ancestry. Kings are in charge of producing the unity of their bounded, internally composite kingdoms by branding everything they contain or take from the outside, with ancestral unifying life substances. Let me expand on this point with regard to the practice of trading.

In the Grassfields, intensive trading in subsistence goods can be documented for the past several centuries at the very least and has been associated with the production of standardised goods. It is most obvious in the case of iron smelting and smithing, the carving of utensils, furniture, musical instruments, and masks in the production of both ordinary and luxury earthenwares, and the manufacture of raphia textiles.

Packaging was extensively used. There were standardised bags, baskets, and calabashes of different sizes for the measurement and trade of grains, palm oil, raphia wine, etc. However, one unusual feature is worth mentioning: The price of certain given goods was the same all over the trading network, but the standardised containers change successively from larger ones to smaller ones as one travelled away from the place of manufacture. Thus, the price of one calabash of palm oil was identical all over the trading network, but the size of a standard calabash diminished as one went further away from the palm-oil producing communities. Smoked fish, caterpillars, grasshoppers, and so on were packaged in stylish and elaborate pieces of basketry. Iron goods were standardised in the form of blanks for hoes or machetes, bound together in bundles of a given number, and traded both as commodities and currencies.

In my view, there is clearly more than a mere analogy between the techniques of the body applied to the human skin as an envelope, the techniques of wrapping and packaging applied to containers and content, and the technologies of power applied to the king and his subjects. For example, producing a successor through a ritual of succession implied anointing the skin of the incumbent and wrapping his body in clothes, herbal medicines, and various adornments (Figure 6.3). The civilisation of the highlands is one of containers, surfaces, openings, and contents, all the way from the kingdom as a bounded body politic down to the baskets, bags, and drinking horns.

The origin of goods and their carriage out of a bounded kingdom, into the bounded space of a marketplace, and all the way into another bounded kingdom and into a given compound was and still is a matter of concern. This is because substances, especially but not only iron objects and medicines, carry power—either nefarious or positive. The origin of a commodity was and still is a problem. It could be tracked down by

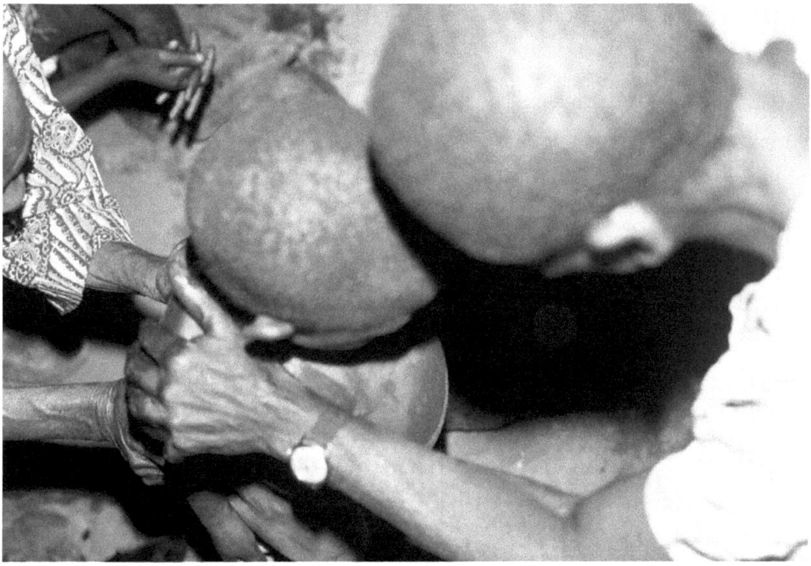

Figure 6.3. Anointing a successor (1973, author's photograph).

reference to the shape and nature of the container (bag, basket, calabash, and so forth specific to a given place of origin) or else by the shape of the commodity itself. The shape of hoes, manufactured in large numbers in several kingdoms specialising in iron production, was different in each kingdom. Another clear example is provided by pottery with recognisable styles. The shapes, wrappings, and styles acted as so many *labels* of origin.

The next question I wish to address is: In such a context, what does branding by the king achieve? What is its effect, or, at least, what do the king and his subjects assume to be its effect? How does it impact on people and things? There is no denying that a brand or a label may be seen as a sign in a system of connotation or communication. It makes sense. However, the underlying assumption behind my question is that it may also be taken as an efficacious action on things and people in a system of agency. In which case, the relevant question is not "What does it mean?" but "What does it achieve? What is its impact? What does it transform and how?"

Let me make an analogy: Branding the skin of a human being with a red-hot iron in the form of a lily or iris (a *fleur-de-lys*) in ancient France certainly conveyed a clear meaning. It imprinted the sign of the crown on the shoulder of the convict. It definitely had a sign value in a coded system of connotation and communication. But it also acted on the person in a system of agency. It transformed that person and directed him (in this example, most commonly male) towards the prison and the galleys with their specific material culture and repertoire of actions. It also prevented the man, should he escape, from acting like any other nonbranded person. Right from the start, this man had been a problem and he had been branded as such and directed towards specific repertoires of action. He had been manufactured as a convict and royal slave (see also Wengrow 2008: 12).

Similarly, the king and his subjects feel concerned about what is in transit through the openings of the king, the palace, the city, and, last but not least, the bodily envelopes of the subjects themselves, and about what is in contact with the envelopes of these various containers. Commodities are a matter of concern.

A good example of such a concern is provided by the imaginary of witchcraft conceived as the action of patronising an occult marketplace to which witches go at night to purchase packaged goods for which they pay by giving away the life of their kin. At the time of purchase, they do not know what is in the parcel. When they bring the parcel back into the kingdom and unpack it, they either discover fantastic riches or things that will cause decay, ruin, and death. In the kingdom of Bum, this occult marketplace is called *msa* (for a more detailed description, see Warnier 1993: 157–62).

This fantasy of an ambiguous market located out there in the occult world illustrates a dichotomy between the outside of the kingdom—made up of the dangerous world at large, where there is a vast diversity and multiplicity of goods and people, labelled in all kinds of ways by all kinds of agents—and the inside of the kingdom in which the king (and the notables) vet and brand both goods and people at the gates of the city (and inside it), thereby achieving a unity of content—a diversified, labelled, external offering met with a unified, branded reception.

Let us go back to the question what does branding achieve under such circumstances? It impacts on the subjects to subjectify them and turn them into the citizens of a kingdom and into the subjects of a given, named, physical king. In other words, it is a political process.

In the past, I assume that the trade in European imports raised specific problems of branding because they came out and from an unknown space and intruded into the kingdom through long-distance trade. They strengthened the political hierarchies but could become a threat to the kingdom and its inhabitants. Accordingly, origins, labels, and branding by the king mattered. They mattered all the more when imported things were close to the skin and to the body, that is, medicines, textiles, ornaments, weapons, gunpowder, etc.

One neat case of a labelled, imported good is provided by guns. The most popular one in the area has been a flintlock manufactured in Birmingham, England, especially for the African market between 1812 and 1840. It was a good-quality musket and many of them are still in use in the early 21st century. This standardised musket bears the proof and view marks of the Birmingham Gunmakers proofhouse (Figure 6.4). It was clearly labelled and was recognised as an authentic version in the Cameroon Grassfields, especially by the inscriptions on the barrel and on the plate, which bear the royal crown of England and the word "tower"—a clear reference to the Tower of London. I have no doubt that other European imports were also labelled, and that the concept of labelled standardised commodities was familiar to the African market before the arrival of the Europeans, paving the way to the success of labelled/branded industrial commodities.

The "Tower" Birmingham flintlock inspired an enduring trust. However, because it was a weapon and, because, furthermore, it was coming from

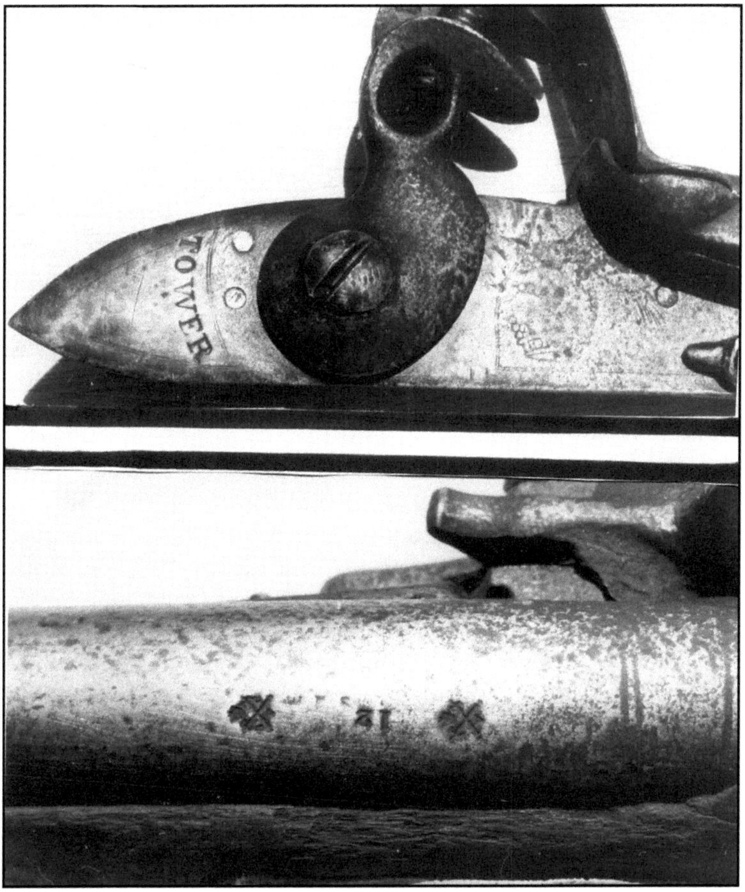

Figure 6.4. The Birmingham flintlock: "TOWER" and crown, proof and view marks (author's photograph).

the outside, the label was not enough. It had to be branded by the king who medicated the guns once a year to remove the threat they represented within the kingdom. This medication is still practised in the 2000s, on the first day of the annual festival. Branding is and was all the more necessary because iron is perceived in the Grassfields as an ambiguous substance: at the same time very powerful and effective and potentially violent and dangerous—much as nuclear energy is perceived in contemporary society (Warnier 2004).

Branding as a Technology of the Body and of the Self

By now, it should be clear that the gestures performed by the king, and the substances he uses, belong to a broad repertoire of techniques of the body shared by himself and all his subjects. The only difference—albeit a major one—is that these techniques, when used by the king, cleanse people and things and achieve their unity within the kingdom. The king is a monarch, that is, a single principle of boundedness and unity. As such techniques apply both to people and material things, they blur the divide between subject and object. People are containers and containers are people (David et al. 1988; Warnier 2005).

If one goes back to the daily massaging of a toddler by his or her mother, it may be seen as a health-related practice, but, in light of the technologies of envelopes implemented by the king, it may also be seen as a daily branding of the child with the double purpose of constructing his skin as a leak-proof, healthy envelope, and of including it into the household of his or her parents and into the kingdom. However, there is more to it because the unifying substances of the king flow out from the palace and cascade over the entire kingdom and its subjects and invest all the material substances used by them. They also brand the commodities entering the kingdom from the outside, including Pears Baby Oil and Lotion manufactured by Unilever in Nigeria for the African market. Thus, the massaging may also be seen by the anthropologist as a means to include the toddler into the kingdom.

To conclude, in such an African context, branding relates to the skin, the body, openings, and envelopes. It is the permanent responsibility of the king, and it concerns both the closure of the kingdom as an envelope and the circulation of things and people in and out of its openings. It is achieved by the use of material royal/ancestral substances and by gestures that, most of the time, are not verbalised and do not elicit any verbal comment or explanation. They belong more to the realm of procedural knowledge and sensori-motricity than to verbalised knowledge. They draw on human motions and emotions regarding the human body and intrusion. They extend to each individual subject and to its concerns with the skin and its treatment, as exemplified by the daily massaging of babies presented in the introduction to this chapter.

Branding may apply to individual nonstandardised things and people. But, it may also apply to standardised commodities. It may be considered as a dialectical and transitional process between outside and inside, diversity and unity, "good" and "bad," the other and the self, the world at large and the kingdom. Branding, in this context, may be performed either to commodify or to decommodify things and people. When an artisan sells something in the marketplace, he may spit on it just before parting with it and handing it over to the buyer. Conversely, the king will anoint a piece of furniture such as a wooden statue with palm oil and crimson camwood to decommodify it and take it as an inalienable possession into the royal treasury.

Notes

1. I am most grateful to David Wengrow and Andrew Bevan for having organised the conference on "Archaeological and Anthropological Approaches to Commodity Branding," to the participants, and to the many people in the Department of Anthropology and the Institute of Archaeology at University College London who contributed to the organisation of the conference and to the discussions. Their questions and contributions helped me rewrite the present chapter for which, however, I assume full responsibility.
2. I have conducted fieldwork in Mankon since 1971. My last stay in the kingdom dates back to September 2002–January 2003. I acknowledge with thanks the financial help of the "UMR-LMS" (CNRS-Université Paris-Descartes).

References

Anzieu, D. 1985. *Le Moi-Peau*. Paris: Dunod.

Bowlby, J. 1958. The nature of the child's tie to mother. *International Journal of Psycho-Analysis* 39:350–73.

Burke, T. 1996. *Lifebuoy Men, Lux Women. Commodification, Consumption, & Cleanliness in Modern Zimbabwe*. London: Leicester University Press.

David, N., J. Sterner, and K. Gavua. 1988. Why are pots decorated? *Current Anthropology* 29:365–89.

Presthold, J. 2008. *Domesticating the World. African Consumerism and the Genealogies of Globalization*. Berkeley: University of California Press.

Rowlands, M. J. 1979. Local and long-distance trade and incipient state formation in the Bamenda Plateau in the late 19th century. *Paideuma* 25:1–19.

Warnier, J-P. 1985. *Echanges, Développement et Hiérarchies dans le Bamenda Pré-Colonial-Cameroon*. Stuttgart, Germany: Franz Steiner Verlag Wiesbaden for the Frobenius Institut.

Warnier, J-P. 1993. *L'esprit d'entreprise au Cameroun*. Paris: Karthala.

Warnier, J-P. 2004. Métallurgie ancienne, identifications et domestication de la violence au Cameroun. In J-F. Bayart and J-P. Warnier (Eds.), *Matière à Politique. Le pouvoir, les Corps et les Choses*: 181–94. Paris/ and Yaoundé, Cameroon: CERI-Karthala and UCAC, coll. "Recherches internationales").

Warnier, J-P. 2005. Inside and outside. Surfaces and containers. In C. Tilley, W. Keane, S. Kuechler, M. Rowlands, and P. Spyer (Eds.), *Handbook of Material Culture*: 186–95. London: Sage.

Warnier, J-P. 2007. *The Pot-King. The Body and Technologies of Power*. Leiden, Netherlands: Brill.

Warnier, J-P. 2009. *Régner au Cameroun. Le Roi-pot*, Paris: CERI-Karthala.

Wengrow, D. 2008. Prehistories of commodity branding. *Current Anthropology* 49:7–34.

Chapter 7

Commodities, Brands, and Village Economies in the Classic Maya Lowlands

Jason Yaeger[1]

Conventional views of commodities in economics define them fairly narrowly in terms of their equivalence of use value. Commodities are characterised as physical substances—often bulk raw materials and staples such as livestock, grains, and metals—that are interchangeable with equal measures of the same substance. A pork belly is a pork belly is a pork belly, at least in terms of its use value. Because its exchange value is subject to supply and demand, real and anticipated, there is also a futures market for pork bellies that have yet to be created. The terms "commodity" and, often more specifically, "petty commodity" are also applied to classes of objects of the same function or use-value, such as hammers, televisions, and axes.

In an economy dominated by commodities, a brand serves several purposes (see Wengrow, this volume). Given that the measures or exemplars of a particular commodity are ostensibly equivalent in use-value, creating a brand is a way to differentiate one subset of that commodity from others. The brand provides a means to make claims about the quality or special properties of the branded commodity. A brand also serves as a vehicle for establishing a special bond between branded goods and consumers (Holt 2004). In contemporary Western business practice, companies use branding strategies in efforts to increase market share and profits.

Anthropologists have tended to understand commodities differently. Igor Kopytoff (1986), for example, defines them quite broadly as any items that have both a use-value and an exchange-value. He suggests there is, in fact, a continuum from pure commodities—objects that are fully exchangeable and substitutable—to inalienable items, objects like gifts that are ostensibly unique and singular. Neither of these polar opposites actually exists in real-world exchanges; each item in every transaction occupies some middle ground. Similarly, Wim van Binsbergen (2005) emphasises the degree to which an object's use- and exchange-values are determined by the social relationships that shape its production and exchange. The weaker their roles, the more an object can be understood to circulate as a commodity.

Moving from individual objects to economic systems, Roy Dilley (2005)—in an approach that I propose to develop here—suggests that we place analytical priority on the degree to which producers and their labour are invisible to the consumer. In social contexts characterised by small-scale regimes of value, such as small village settings, a consumer's knowledge of the processes of production and his or her personal relationships with the producer both strongly influence the negotiation of an exchange value for an object. This is particularly true if objects are obtained directly from the producers, without mediating vendors. In these village economies, objects and their values are strongly structured by local social relations.

If, however, the context of production is divorced from the point of purchase and context of consumption, the seller has more latitude in setting the terms of exchange. In the bazaar economy as formulated by Frank Fanselow (1990; also Wengrow 2008, this volume), brokers acquire goods of uneven quality from producers and offer them for resale to consumers or to vendors who then resell them. The point of purchase and context of consumption are thus spatially and socially removed from the loci of production and from the producers. This creates a strong asymmetry in knowledge about how an item was produced and its quality, providing the vendor some advantages in negotiating the item's exchange value. Of course, the consumer can seek to counteract this asymmetry by shrewd bargaining and comparison shopping with multiple vendors. Finally, in an economy dominated by branded commodities (Wengrow, this volume), brands serve to distinguish amongst the various examples of a given commodity. Producers seek to maintain a certain level of quality and consistency in their branded goods; thus, the brand holds a promise of particular characteristics and qualities. In this way, brands can reduce consumer anxiety of buying goods of variable quality that is inherent in the bazaar economy.

These anthropological perspectives emphasise processes of exchange and valuation and they employ the construct of the commodity as a useful abstraction. No item is inherently a commodity; rather, items act more or less like commodities in particular exchanges. These shifts in emphasis focus our attention on the following questions: What structural conditions encourage the commodification of goods? What conditions foster the establishment of brands? To address these questions, I will consider the production and exchange of pottery in a Classic Maya economy. I draw my data from the neighbouring polities of Xunantunich and Buenavista del Cayo, Belize (LeCount et al. 2002). I have chosen to focus narrowly on two centres in the Mopan Valley because a close analysis of the data from these two sites is the best means to elucidate the relationships that existed between consumers and producers of certain goods. Even so, given the social, political, and economic heterogeneity of Classic Maya civilization (discussed below), I remain cautious about overgeneralising my conclusions to other Maya polities.

Classic Maya Civilization

The central and southern Maya lowlands were occupied by dozens of competing polities during the Classic period (AD 250–900). These polities were small in size, with populations ranging between 15,000 and 75,000 people, and closely spaced, often less than 15 km apart (Demarest 2004; Sharer 2004; Webster 1997). The larger polities were ruled by a *k'uhul ajaw* ("divine lord"; pl. *k'uhul ajawtak*), a title that implies a high degree of political autonomy (Sharer and Golden 2004) and is sometimes translated as "king" (Martin and Grube 2008).

Maya *k'uhul ajawtak* drew legitimacy from military successes and from their preeminent roles as intermediaries with the supernatural, celebrating both of these types of activities in the public art they commissioned (Sharer and Golden 2004). They also sought to maintain their political preeminence by creating and maintaining bonds of loyalty with their subordinates through gift giving (Ball 1993a; LeCount 1999). Maya rulers maintained craft workshops where artisans made elaborate, high-value objects, including ceramic vessels and jewellery, which the rulers then distributed to their followers and allies (Ball 1993a; Foias 2002; Inomata 2001). These gifts materialised the unequal interpersonal bonds between the *k'uhul ajaw*, as disperser of largesse, and his or her subjects, the subordinated recipients. Because these subjects lacked the resources to make comparable objects, they reciprocated with their loyalty and whatever services they could offer their sovereign (Foias 2002).

Maya centres were not densely nucleated cities, and their widely spaced residential complexes were separated by terraces and other agricultural features (Pyburn 2008; Webster 1997). Population densities generally declined little as one moved outwards from the polity's centre into the surrounding countryside (Drennan 1988; Yaeger 2003), where farmers practiced a range of intensive and extensive agricultural strategies.

The countryside was also the scene of significant craft production. Most farming households were also crafting households, engaging in spinning, weaving, flint-knapping, potting, and other craft activities (Pyburn 2008; also Hirth 2009). Other hinterland residents dedicated themselves to more intensive and specialised craft production, such as the manufacture of fine bifacial stone tools and, probably, the production of pottery, as detailed below.

The ways through which these goods circulated within Maya economies are the subject of much debate (see Wells 2006). Many Mayanists accept a dual model of the economy that draws a strong distinction between prestige goods and utilitarian goods (e.g., Foias 2002). In this model, prestige goods were manufactured under elite patronage, often in attached workshops. Some of these items were reserved for use in ceremonies and rituals; others were distributed by the elite as gifts to their subjects. In contrast, the manufacture of utilitarian goods was the purview of relatively autonomous artisans who worked at or near their rural homes (Sanders and Price 1968; Webster et al. 2000). Some scholars have inferred that the producers of utilitarian goods exchanged most of their wares with members of their kin group or community (Sanders 1989). Others have argued that marketplaces existed to facilitate exchange (e.g., Jones 1996), a conclusion supported by a growing body of empirical evidence as discussed below.

Injecting the constructs of commodity and brand into this debate draws our attention more closely to the complex triadic relationships between producers, consumers, and goods they exchanged. Notions of brand and commodity bring production, exchange, and consumption together into a coherent dynamic that is inherently social, and that does not rely on a top-down model of the political economy, although it is not incompatible with top-down perspectives. The construct of brand in particular forces us to consider the recursive dynamic between consumer decisions and the strategies of producers to influence those decisions and of the social relationships forged by that dynamic. Thus, looking at Classic Maya economies through the lenses of commodity and brand compels us to shift from thinking about the Maya economy as an awkward amalgam of prestige goods that circulate through politically generative gift exchanges and utilitarian goods obtained in strategies of household provisioning, emphasizing exchange and consumption as socially situated practices instead.

The Mopan Valley during the Classic Period

The Mopan Valley of western Belize presents a broad and fertile alluvial floodplain framed by gentle, karst-derived limestone hills. The hills are carpeted with agricultural terraces built by farming families over generations as they transformed the valley's landscape for intensive agricultural production (Neff 2008; Robin 1999; Wyatt 2008). In the floodplain and hills, deposits of clays and chert nodules provided important raw materials for Classic Maya potters and flint-knappers (Yaeger 2010). The river itself was a natural transportation corridor, connecting the landlocked areas of the central lowlands, home to powerful Classic polities like Naranjo and Tikal, to the Caribbean Sea and coastal trade routes (Figure 7.1).

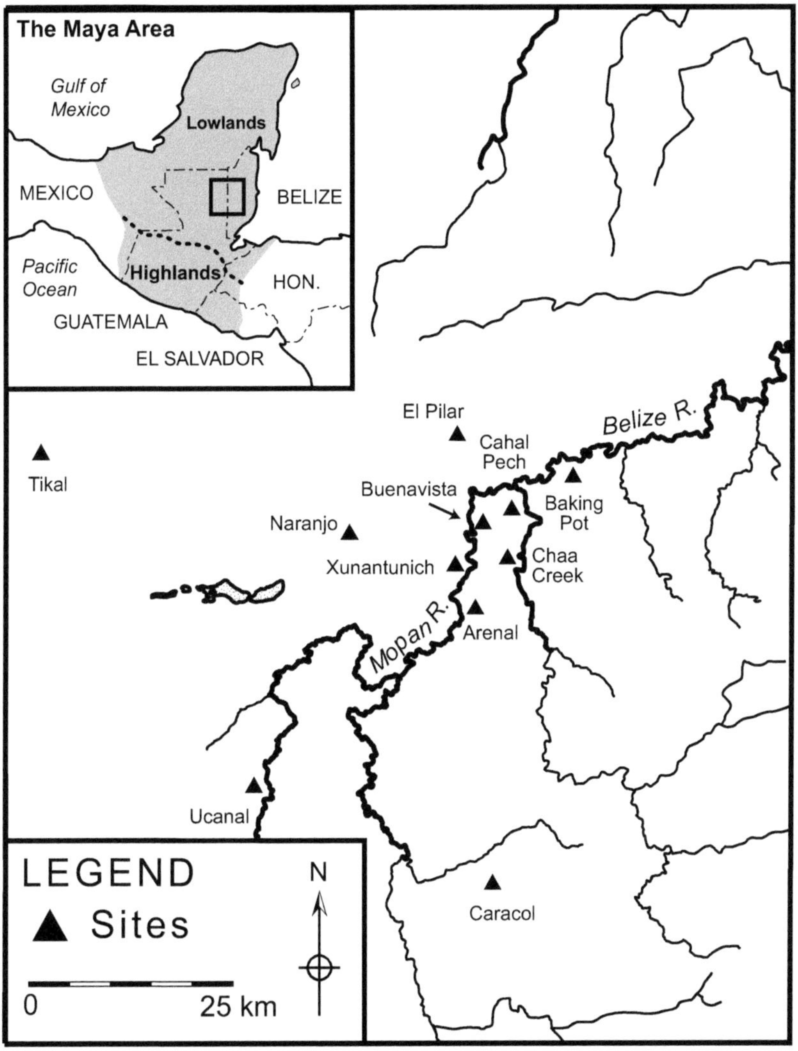

Figure 7.1. Archaeological sites of the Mopan Valley.

During the Classic period (AD 250–900), the Mopan Valley was densely packed with political centres, and Richard Leventhal and Wendy Ashmore (2004; also Ashmore 2010; Helmke and Awe 2008) have suggested that this close spacing was the result of the sequential rise of competing political centres over the course of the Preclassic and Classic periods. Buenavista del Cayo (hereafter Buenavista) was founded around 950 BC and developed into one of the valley's most important political centres beginning around 650 BC (Ball and Tascheck 2004). Xunantunich would not be established as a major political centre for another millennium (LeCount and Yaeger 2010a). By the early part of the Late Classic period (AD 600–670, the Samal phase), however, Buenavista and Xunantunich were the two dominant centres in the Mopan Valley. These two centres competed for the valley's resources, including the loyalty of the valley's populace (LeCount and Yaeger 2010b; Leventhal and Ashmore 2004; compare Taschek and Ball 2004), and Xunantunich emerged as the most powerful centre in the valley by the later Late Classic period (AD 670–780; Hats' Chaak phase). The Terminal Classic period (AD 780–890; Tsak' phase) saw diminished construction in most of the larger centres and the abandonment of all of them by the end of the period.

The Mopan Valley countryside was densely populated, as documented by systematic settlement surveys across the region (Ashmore 1995; Ford and Fedick 1992; Yaeger 2010). The population in Xunantunich's hinterland reached its peak of more than 400 people/km^2 during the late 7th and early 8th centuries AD (Neff 2010; Yaeger 2003). The countryside was organised in a complex manner, with elite residences that were the homes of wealthy rural families; administrative complexes used by polity rulers to project their authority into the hinterland; and hamlet-sized settlement clusters of variable longevity, some of which show remarkable socioeconomic heterogeneity for their small size (Yaeger and Robin 2004). The massive Late Classic construction projects at Xunantunich and other centres testify to the authority that the region's rulers exercised over hinterland communities, and survey and excavation data from the hinterlands also show that Xunantunich's rulers strategically engaged with preexisting hinterland social and political networks (Ashmore et al. 2004). At the same time, these strategies were neither wholly effective nor uniform, and some hinterland groups grew stronger through the new arrangements that emerged out of their negotiations with regional rulers (Robin et al. 2010; Yaeger 2000a, 2003).

This highly diversified countryside was the venue for significant economic production and consumption. Rural families not only remodelled the landscape over generations to increase agricultural yields, but also engaged in other productive activities, presumably in efforts to generate wealth and disposable income, meet tribute demands, and buffer themselves against the risk of poor harvests. At least a few made ornaments from imported marine conch shells (Yaeger 2000b); most spun cotton and likely wove it into cloth (Robin 1999; Yaeger 2000b).

Markets and Marketed Goods in the Late Classic Mopan Valley

Stone tool production provides us with our best starting point for assessing the region's economic complexity. Chipped chert tools and associated waste (debitage) from their production is common in domestic Maya household middens and demonstrates that most farming families produced some simple stone tools, likely for personal consumption (VandenBosch 1999).

There was also a second, more specialised level of manufacturing, in which one or more households formed a workshop. The extremely high densities of debitage and tools in these workshops (e.g., 1–2 million flakes and 14 finished bifaces per cubic metre at T/A1-002 and T/A1-003; see VandenBosch 1999) speak to the great numbers of tools they produced. Workshops manufactured more complex tools like spear points and axes, and different workshops seem to have specialised in different subsets of tools. This likely reflects the varying quality of local stone outcrops, but could also reflect a concern with enhancing the desirability of particular tool types within and beyond the community (e.g., Yaeger 2000b). There were at least three workshops within 4 km of Xunantunich (Hearth 2008; VandenBosch 1999; Yaeger 2000a), demonstrating that there were multiple closely spaced loci of intensive production of these essential goods, which likely competed with one another for consumers.

A third production pattern occurs at the major centres of Xunantunich and Buenavista themselves, in large plazas that likely served as central marketplaces. Although some scholars have argued for the existence of marketplaces in Maya cities for decades (e.g., Jones 1996; Sabloff and Tourtellot 1992), it is only recently that archaeologists have begun combining architectural patterns, artefact distributions, and soil chemistry to more confidently identify them (Cap 2007, 2008; Dahlin 2003; Dahlin and Ardren 2002; Keller 2006; Wurtzburg 1991). Although I will not review the evidence in detail here, two locations identified as markets were also venues for stone tool production: the Lost Plaza at Xunantunich (Keller 2006) and the East Plaza at Buenavista (Ball 1993b; Cap 2007, 2008; Heindel 2008; Reith 2003). The evidence at the latter site is particularly compelling; flint-knappers' efforts were largely restricted to the very last steps in the production process, putting the finishing touches on nearly completed tools and resharpening tools dulled by use (Heindel 2008). This is consistent with final-stage, point-of-sale production by manufacturers who are also vendors, a pattern also suggested by Keller (2006) at Xunantunich.

The identification of marketplaces at Xunantunich and Buenavista adds a critical component to our understanding of the Mopan Valley's economy. A centrally located marketplace is an efficient mechanism to meet consumer demand in a large population (Beals 1975). Furthermore, as Pyburn (2008) has argued, marketplaces were probably important sources of income for the rulers and elites who governed Maya cities, as regulating and taxing exchange and distribution is much more efficient than attempting to directly control production. Turning to the impacts of marketplaces on consumers and exchange relations, marketplaces remove the point of exchange and thus the consumer from the locus of production. In that way, marketplaces can impede consumers' ready evaluation of the resources (time, energy, raw materials, expertise) needed to produce a particular object and limit their ability to use those factors in negotiating an exchange value for that object. Consequently, sellers are in a stronger position to set the terms of exchange and make claims of value that are disassociated from production technology and disentangled from many local social relations. At least for many stone tools at Xunantunich and Buenavista, however, producers and consumers engaged directly at the point of exchange, without intermediaries, as stone tool vendors also made tools.

While marketplaces were important at Buenavista and Xunantunich, it is implausible that they were the only venues for exchange in the region. Some transactions occurred between neighbours and families (VandenBosch et al. 2010), whereas others probably involved itinerant merchants. The goods exchanged at marketplaces likely included objects made of raw materials imported from long distances such as obsidian (Cap 2008; Keller 2006) and marine shell; regional imports like granite grinding stones for daily food preparation (Abramiuk and Meurer 2006; Graham 1987), pine for fuel (Lentz et al. 2005), meat (Yaeger and Freiwald 2009), and probably salt (McKillop 2002); and locally made objects like chert stone tools, cloth, basketry, and pottery. Of all classes of goods, however, chert tools and pottery vessels were probably the most commonly exchanged objects in Maya marketplaces, although we still lack conclusive evidence of the latter.

Ceramic Production and Consumption in the Late Classic Mopan Valley

Our understanding of pottery production in the Mopan Valley is largely derived from stylistic and formal analyses, coupled with chemical and petrographic characterisation. Despite a long history of research in the valley, archaeologists have identified only one locus of pottery production, a workshop for painting elaborate polychrome vessels located in an elite residential complex at Buenavista (Ball 1993a; Reents-Budet et al. 2000). Although many critical facets of the valley's ceramic production systems remain poorly understood or entirely unknown, we are justified (on the basis of evidence for other industries; see above) in considering whether consumer demand had a significant role to play in shaping the activities of local pottery makers. I will elaborate on this below, through the examination of three classes of ceramic vessels from the Late Classic Hats' Chaak phase (AD 670–780). I have selected these types because they span the range from the most elaborate vessel type to one of the simplest types (following LeCount 1996).

One important issue relevant to the question of commodities and brands is whether consumers had access to functionally equivalent vessels of distinct kinds, regardless of whether they were acquired through marketplace exchange or some other means. How many choices did consumers have, and thus what was the potential role for branding to differentiate amongst commodities and shape consumer choice? In the case of some of the most widely used pottery forms, utilitarian black-slipped bowls and jars of the Mount Maloney Black type, discussed below, consumers could and did acquire similar items made by different workshops.

This raises the possibility of thinking of particular combinations of visually distinct stylistic and/or formal attributes as a brand. As understood by most economists, commodity branding is a process for differentiating amongst goods that have similar use values and thus are amenable to price-based consumer decision-making. From the producer's perspective, there are three keys to successful branding: The brand must be distinctive and recognisable; the brand must be controllable; and the brand must hold some unique promise of value. By branding, a producer differentiates his or her product and adds some perceived value that forms the basis of a special bond between consumers and the branded goods. Thus, the brand motivates consumers to favour that particular brand over alternatives and allows producers to capture market share.

From the perspective of the consumer, brands have several purposes. Wengrow (2008) emphasises the ways that brands serve to guarantee that a commodity has particular qualities, the promise of value just mentioned. They can also be important for mediating social relationships (Bevan, this volume), and consuming brands also can be powerful statements of identity. In what follows, I will discuss three distinct classes of Maya ceramics and consider whether they can be understood productively as branded goods. Much of the detailed evidence I present may be of interest primarily to specialists, but is nevertheless essential background to my wider conclusions regarding the nature of commodity production and circulation in the Classic Maya economy.

Imported Painted Serving Vessels

Amongst the best-known products of Classic-period Maya artisans are polychrome vessels painted with complex narrative scenes (e.g., Figure 7.2). Many of these are tall cylindrical vases, which often bear hieroglyphic texts that name them as vessels for drinking *kakaw,* a chocolate beverage (Stuart 2006). In fact, this function has been verified through residue analysis (Hall et al. 1990). Other common forms include smaller bowls and large platters, for drinks like *sa'* (a corn-based beverage) and for serving food like *wah* (*tamales*), respectively (Houston et al. 1989). Most scholars believe that these fine vessels were used by the royalty and highest elite of Maya society for consuming food, particularly in feasts and other ceremonial meals (Reents-Budet 2000). The close association between noble individuals and these vessels is confirmed by their placement in the tombs of elite individuals and the texts on vessels that name them as belonging to particular high-status individuals (e.g., Figure 7.2).

These vessels were not likely produced in large numbers—they are exceedingly rare, and each one would have required a considerable investment of time, specialised knowledge of symbolism and hieroglyphic writing, and artistic expertise. Painters likely practiced for years to hone their skills (Reents-Budet 1994). Although we have not identified with certainty a single workshop for the production of these fine vessels, most scholars believe they were produced in palace schools, workshops attached to and controlled by the royal families of Maya polities (Ball 1993a). They are unlikely to have been exchanged frequently, and when they were, the exchange likely entailed the ruler who controlled the palace workshop bestowing one as a gift to a vassal and ally, whether existing or prospective. Acceptance of the gift acknowledged a bond between the receiver and the giver and presumably entailed an expectation of reciprocation on the part of the recipient. Thus, they helped constitute far-reaching, inter-elite hierarchical political networks that could be mobilized in strategies of territorial expansion and the control of long-distance trade. The use of

Figure 7.2. The Buenavista Vase (Rollout photograph, K4464 © Justin Kerr, 1992, with permission).

these vessels in ceremonial meals made the social and political bonds that constituted those networks materially manifest to all participants.

Unfortunately, because of their value on the antiquities market, the majority of these vessels available for study today are unprovenanced pieces held in museums and private collections. Despite their lack of provenience, art historians have identified different schools of polychrome vessel painting, each with distinct canons of iconography, narrative, painting style, and colour schemes. Chemical charcterisation of the clay used in vessels has allowed archaeologists (e.g., Reents-Budet 1994) to determine the locations of many schools. This, in turn, has allowed archaeologists to demonstrate that vessels sometimes made their ways into the hands of individuals in polities far from their point of production, likely through inter-elite gifting.

One of the best examples of this phenomenon was documented at Buenavista. There, Jennifer Taschek and Joseph Ball (1992) discovered a noble tomb that contained, amongst other objects, a Cabrito Cream-Polychrome vase with a hieroglyphic text (Figure 7.2). Although it is called the Buenavista Vase, evidence suggests it was made in a palace workshop in the much larger kingdom of Naranjo, located some 15 km to the west. Instrumental neutron activation analysis (INAA) of the vessel's clay raw material (Reents-Budet and Bishop 2003), the style in which it is painted (Reents-Budet 1994), and the text it bears (Houston et al. 1992) all point to the object's origin at Naranjo.

The scene on the Buenavista Vase shows the so-called Holmul Dancer, a depiction of a nobleman dancing as the Maize God, often wearing an elaborate back-rack and accompanied by dwarfs (Reents-Budet 1991; Taube 1985). The text states that the vase was the drinking vessel of the Naranjo's powerful and expansionary *k'uhul ajaw*, K'ahk' Tiliw Chan Chaak (Houston et al. 1992). During his reign (AD 693 to at least 728; Martin and Grube 2008), he built alliances and pursued conquest to extend his influence over neighbouring polities in the eastern Peten and the Mopan Valley (Helmke and Awe 2008; Houston et al. 1992; LeCount and Yaeger 2010b; Martin and Grube 2008; Taschek and Ball 1992). K'ahk' Tiliw Chan Chaak likely gave one of his personal drinking vessels to the ruling family of Buenavista in his efforts to extend his kingdom's influence into the Mopan Valley (Houston et al. 1992; Taschek and Ball 1992).

Vessels like the Buenavista Vase bear little resemblance to commodities. As noted, they often bear texts naming their original owner, and they were important heirlooms, buried with elite individuals. These facts indicate an intimate connection between owner and object. Furthermore, their circulation as gifts in politically charged exchanges attached values and meanings to them that were densely entangled in the social and political relationships that structured their gifting. These relationships also would have inhered in the use of those vessels. The concept of commodity applies to these elaborate and rare polychrome vessels as well as it does to Fabergé eggs. Indeed, it is only within the vastly different context of 20th-century capitalism that Maya polychromes have become commodified to some extent through the art and antiquities market.

The Naranjo school of Cabrito Cream-Polychrome vessels does have much in common with contemporary brands, however (see Bennet [2008] for a similar example from the Aegean). The vessels from this particular palace school were distinctive and recognisable; their production within a palace workshop context would have allowed for close control of their distinctive features, analogous to brand management; and the social ties to the royal house of Naranjo that those features implied constituted a unique promise of value to their users. Furthermore, they were used in socially generative practices, during which people plausibly expressed important facets of their self-identity (such as elite status and sophistication) and represented their connections to the individual who had given them the gift. Inasmuch as consumer choice is a productive construct in the exchange of these vessels, such choice was influenced strongly by the *realpolitik* calculus of forging alliances with the rulers of Naranjo.

Locally Made Painted Serving Vessels

At Xunantunich and Buenavista, local potters made serving vessels that were much more widely distributed than those just discussed. Some of these, like the large plates and dishes of the monochrome Belize Red group, are common across the entire upper Belize River Valley (Gifford 1976; LeCount 1996). Others are more restricted in their distribution. At Xunantunich and in adjacent hinterland sites, vessels with relatively simple designs painted on an orange background predominate; at Buenavista, vessels with designs painted on a distinct cream background are more common. A comprehensive study of cream-slipped painted pottery recovered at Buenavista that combined detailed stylistic and formal analyses by Joseph Ball with INAA sourcing by Dorie Reents-Budet and Ronald Bishop revealed several distinct ceramic groups, some of which were produced locally at Buenavista (Reents-Budet et al. 2000).

A relatively simple painted pottery type produced at Buenavista is called Chinos Black-on-Cream. These vessels tend to be small flat-bottomed dishes with outflaring walls and small, straight-sided vases (Reents-Budet et al. 2000: 103). The most common decorative motifs are relatively simple *k'in* solar or flower symbols and stylised representations of insects, painted in black on a cream background. The Chinos Black-on-Cream type presents strong iconographic and formal similarities to the Chunhuitz Orange group that is ubiquitous at Xunantunich, in which *k'in* signs are the most common motif and the flat-bottomed outflaring dish is the most common shape. Despite these shared features, the distinctive cream background of Chinos Black-on-Cream sets them apart from Chunhuitz Orange.

A locally produced pottery type more elaborate than Chinos Black-on-Cream is the Group 1a subtype of Cabrito Cream-Polychrome (Reents-Budet et al. 2000). Many of these vessels are barrel-shaped vases, and they often bear figural scenes such as the Holmul Dancer, rendered in orange and red paint on a cream background. The figures are sometimes separated by vertical bands of motifs or pseudoglyphs, elements taken from hieroglyphic writing that do not form coherent texts. A band of pseudoglyphs runs around the exterior rim of many vessels, as well, in the same area where fancier vessels often bear a true hieroglyphic text.

Group 1a shares many characteristics with the finer varieties of Cabrito Cream-Polychrome produced in the palace school of Naranjo: the colour scheme, the Holmul Dancer theme, and the use of bands of glyphs. They tend to be poorer in execution and attention to detail, however, lacking the complexity of iconography, the fine control of line and colour gradations, and legible texts (Reents-Budet et al. 2000: 110). They could be seen in some ways as imitations or knock-offs of the fancier Cabrito Cream-Polychrome vessels made in the Naranjo school (but see Pinheiro-Machado, this volume, for a cautionary tale against simplistic assumptions about the consumption of knock-off brands).

The distribution of Group 1a throughout the Buenavista polity has not been fully published, but Ball (1993a; also Reents-Budet et al. 2000) has suggested that the most elaborate vessels were used by the elite for food and beverage service and served as social currency in the local political economy. Given their restricted distribution and high value, Ball (1993a; also Reents-Budet et al. 2000) has argued that these types were produced in one or more palace schools controlled by the royal family of Buenavista, perhaps even located within the palace itself (Reents-Budet et al. 2000: 113). Thus, in Ball's model, these vessels not only parallel the finer Naranjo Cabrito Cream-Polychrome type in their design and execution, but in their production, distribution, and consumption as well: they were produced in attached workshops, distributed through elite gifting networks, and used in socially diacritical elaborate meals.

Ball (1993a: 258) has argued that the distribution of vessels produced in Buenavista's palace schools should define the limits of the political reach of the rulers who controlled that workshop. More recently, Ball and Taschek (2004: 159–60) have argued that a decorative motif that they term the Buenavista Device appears on locally produced cream-slipped painted pottery. They tentatively suggest that this symbol refers explicitly to the Buenavista polity or its ruling house and mention a possible Cahal Pech Device, emblematic of that nearby polity (Ball and Taschek 2004: 165).

There are two additional locally made cream-slipped ceramic groups (Groups 1b and 3) that are closely related to Group 1a stylistically and chemically but are somewhat cruder in their manufacture and the execution of the designs painted on them. Although Ball and his colleagues suggest they were produced in the same palace school as the finer Group 1a vessels (Reents-Budet et al. 2000: 114), these features, coupled with a wider range of forms, lead them to argue that vessels of this group were used for daily food service in both high- and low-status elite households, and perhaps even in nonelite households.

At Xunantunich, cream-slipped pottery is rare after *ca.* AD 670, and the vast majority of the painted pottery belongs to the Chunhuitz Orange group (LeCount 1996). The most common vessel shapes in this group are serving forms: plates, dishes, and bowls with flaring or slightly outcurving sides; barrel-shaped vases and incurving bowls are also common (LeCount 1996; also Gifford 1976). Lisa LeCount (2001: 946; also 1996) uses ethnographic analogy to argue that small bowls and dishes were used for individual food service in both sacramental and quotidian contexts and vases were used to drink chocolate and other beverages in those same contexts. Chunhuitz Orange group vessels have a matte orange background, created either by the polished surface of the vessel's volcanic ash-tempered body or a thin orange wash. Although they are occasionally fluted, application of red- and black-painted motifs is the ubiquitous decorative technique within this group (Figure 7.3a).

Unfortunately, the volcanic ash temper makes these vessels very friable, and they are often recovered in very eroded condition, making systematic and detailed analysis of the painted motifs challenging. Nonetheless, painted triangles, *k'in* symbols, and wavy E and S forms are most common, usually found on the exteriors of dishes and bowls (LeCount 1996: 418, 1999: 249; Thompson 1942). Some vessels clearly were more complex in their decoration, bearing multiple geometric motifs and very rarely a band of hieroglyphic text or pseudoglyphs around the rim. Figural representations of animals and people are almost unknown, in stark contrast to the Cream-Polychrome tradition documented at nearby Buenavista.

The Chunhuitz Orange group vessel shapes suggest they were predominantly serving vessels, and their distribution indicates they were used by many within the Xunantunich polity. Their distribution, however, is uneven. In the fullest analysis of the distribution of ceramic types in the Xunantunich polity, LeCount (1999: 252) tentatively suggests fine vessels like those of the Chunhuitz Orange group owed their distribution to gifting networks within a prestige-oriented political economy. Thus, she interprets their increasing scarcity as one moves from elite contexts at Xunantunich to hinterland elite households and finally to

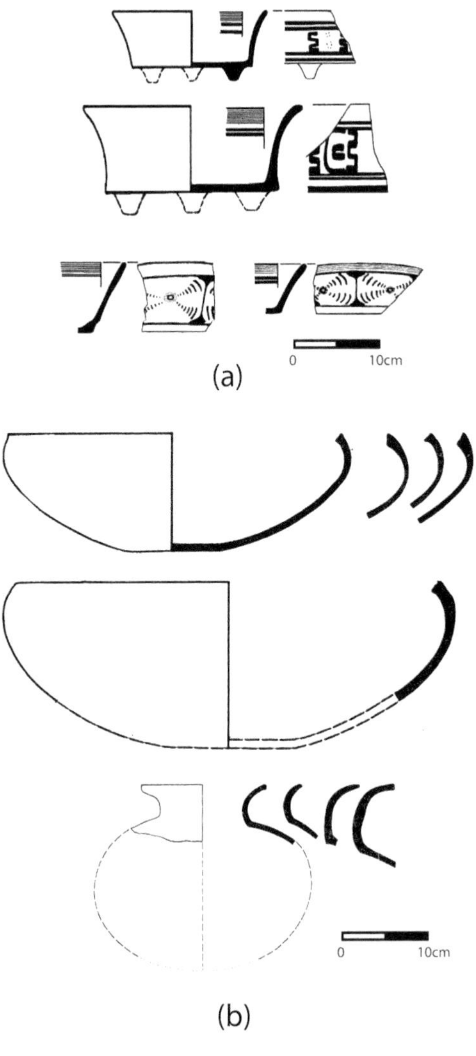

Figure 7.3. (a) Benque Viejo Polychrome bowls of Chunhuitz Orange group (after Thompson 1942: Figure 35); (b) Mount Maloney Black bowls and restricted jars (after Thompson 1942: Figures 26 and 28).

nonelite hinterland households as evidence of the increasingly marginalised positions of these groups within Xunantunich's political economy. This conclusion is in line with Ball's interpretation of the finer Cream-Polychromes at Buenavista, discussed above. Alternatively, these more elaborate vessels may have been more costly to make (demonstrated by LeCount [1996, 1999] and suggested explicitly in LeCount 2001: 947), and thus the fall-off in their distribution as one moves to nonelite households could be explained by differences in disposable household wealth. LeCount (2001) has also argued that the presence of relatively few small bowls and vases in commoner households suggests a reduced emphasis on feasting and related public consumption events (see also Yaeger 2000a).

I find it quite plausible that both mechanisms suggested by LeCount for the distribution of Chunhuitz Orange vessels were operating simultaneously in the Mopan Valley. It is likely that these vessels could have been acquired at the Xunantunich market by those with the means and desire to do so. At the same time, their elaborate decoration and relative high cost, coupled with their highly visible role in feasts and celebrations, made them appropriate gift items, as their later use in meals would recall and evoke the original gift. I further suggest that their small size would have made them ideal items to be given as mementos during elite-sponsored feasts and public ceremonies at Xunantunich. Bishop Diego de Landa (Tozzer 1941:27) documented the central importance of gift-giving during ceremonies and feasts in Maya civilization at the time of Spanish Contact (Helmke 2001; Reents-Budet 2000, 2006; see also Vaughn [2004] for a South American example).

Locally Made Utilitarian Vessels

Serving vessels like those just described were complemented by utilitarian vessels used for food preparation and storage. The most common ceramic type in the Mopan Valley around Xunantunich is Mount Maloney Black (Figure 7.3b). These ubiquitous black-slipped vessels constitute between 38 and 47 percent of the Late and Terminal Classic ceramic assemblages (LeCount 2010b). Given this, it is not surprising that they were made with local clays and tempers (Garcia 2008). Although they were fashioned in a range of shapes, the two most common forms are deep, incurving bowls (Figure 7.3b) and restricted-neck jars (Figure 7.3b), both of which are ubiquitous in household assemblages (LeCount 1999) and common in burials and ritual contexts (LeCount 2010a). LeCount (1996, 2010b) has argued that the deep bowls were used to prepare dough for maize-based foods like *tamales* that were the staple of the Classic Maya diet, and that the restricted jars were used for carrying and storing water. Although comparable statistics have not yet been published, vessels of the Mount Maloney Black group predominate in domestic assemblages and are found in burials and ritual deposits at the large centres immediately surrounding Xunantunich including Buenavista and Arenal, and at smaller sites as well, like Nohoch Ek (Taschek and Ball 1992, 1999, 2003).

Given the relatively high degree of standardisation of both their formal and stylistic attributes (LeCount 1999) and the clay and temper combinations that comprise their ceramic pastes (Garcia 2008), it seems likely that the region's Mount Maloney Black vessels were produced by household- or community-scale workshops. Petrographic analysis of thin-sections of vessels from Xunantunich and two hinterland sites revealed four discrete paste recipes, all of which endured from the early 7th century through the late 9th century AD (Garcia 2008), suggesting the presence of four long-lived units of production, probably four distinct workshops. Garcia's analysis suggests that the same workshops made Mount Maloney Black vessels and Cayo Unslipped storage jars, supporting earlier models that potting workshops made a variety of utilitarian vessel forms in several distinct ware categories (Ball 1993a: 245).

The presence at three widely dispersed sites of vessels made with each of the four distinct paste recipes suggests that vessels from these four workshops were exchanged at a central location, likely the marketplaces at Xunantunich. There is no evidence that these workshops operated under any kind of elite supervision or control, although we have not identified any production locales to date.

As one moves east down the valley from Buenavista and Xunantunich towards Cahal Pech and beyond to Baking Pot and Barton Ramie, Mount Maloney Black bowls and jars become increasingly uncommon, and they are replaced by vessels of the same basic shape classes that belong to the Garbutt Creek Red and Vaca Falls Red types. These red-ware types are visually differentiated from Mount Maloney Black types by their red slips, body shapes, and distinct rim and lip treatments (Gifford 1976; LeCount 1996). These red wares presumably were produced in workshops in the eastern reaches of the valley, where they dominate the ceramic assemblage in the same way that Mount Maloney Black does around Xunantunich and Buenavista.

Aimee Preziosi (2003) and Sam Connell (2003, 2010) have argued that Mount Maloney Black pottery was emblematic of a political identity tied to Xunantunich. At the hinterland site of Chaa Creek, Connell has shown that Mount Maloney Black vessels became more popular than red-slipped vessels in both nonelite and elite households as Xunantunich rose to a position of political dominance in the region during the late 7th and early 8th centuries AD. The ratio of black to red vessels shifted from approximately 1:1 to 3:1 during this time (Connell 2010). This shift is most pronounced amongst the site's elite leaders, leading Connell to argue that they were more actively expressing political affiliation to the region's new dominant power as part of their strategies to maintain their place within the region's shifting power structures. In contrast to this instrumentalist interpretation of the roles of Mount Maloney Black pottery in practices of political identification, LeCount (2010a, 2010b) has argued that Mount Maloney Black vessels were key material components of quotidian bodily practices associated with preparing maize and carrying water, and that they thus contributed to a collective and embodied community identity in the region.

A third explanation, not incompatible with the others, is that the rise of Xunantunich late in the 7th century AD drew more people to the site and its marketplace and that people from Chaa Creek more frequently travelled to Xunantunich's marketplace, where they obtained black-slipped vessels, as opposed to the red-slipped vessels that were more readily available at Cahal Pech and other centres to the east. Elite families might have had more reason to travel to the new centre of political power, as they sought to negotiate a favourable position for themselves as regional power structures shifted during the competition between Xunantunich and Buenavista.

The black slip and lip and rim attributes of Mount Maloney Black vessels set them apart from other contemporaneous red-ware groups of the same vessel classes. This differentiation evokes the way a brand serves to distinguish one subset of a given commodity from other subsets of that commodity. It is not clear, however, that these red-slipped and black-slipped alternatives were widely available in the same markets. Furthermore, when we turn to the four Mount Maloney Black workshops identified near Xunantunich, we find no evidence of branding. Potters obviously controlled their recipes—perhaps even guarded them jealously—and passed them down over generations, maintaining a high degree of product consistency, but they did not seek to create strong visual distinctions between the vessels produced by different workshops. Instead, the most obvious differences in visible attributes relate to change over time (Garcia 2008; LeCount 1996), not unit of production (compare Taschek and Ball 2003: 378). It is difficult to assess whether the standard slip colour and vessel forms and lip treatments employed by multiple workshops were the result of an agreement amongst those workshops to manufacture a standard and distinct black-slipped product or simply the local customary way of making pottery.

Although more research is required to identify the loci of production and distribution of red- and black-slipped vessels, at this juncture it seems most likely that the factors governing the acquisition of Mount Maloney Black vessels were the proximity of producers and consumers to the valley's different marketplaces and the social, economic, and political factors that drew consumers to the sites with those marketplaces. If this was the case, then, as with the stone tool economy (see above), exchange was distanced from the loci of pottery production, allowing vendors to more effectively set the terms of exchange and claim special properties and promises of value for their wares.

Discussion and Conclusion

Although the data currently available provide an imperfect and fractured understanding of the economic organization of the Mopan Valley, they do permit some tentative inferences pertinent to the themes addressed in this volume. One of the questions raised in this volume is whether commodity branding is a phenomenon unique to capitalist economies (also Wengrow 2008). This case study suggests that branded commodities like those found in today's branded economy did not exist in Late Classic Maya civilization.

Some goods certainly could be considered commodities, but they were not clearly branded goods. In the case of both chipped stone tools and utilitarian pottery, multiple contemporary workshops produced the same kinds of objects, such as chert axes and ceramic jars and bowls. Potters and flint-knappers tailored their production decisions in light of the characteristics of the raw materials they used, and they certainly paid attention to consumer needs and desires. That said, at least in terms of pottery for everyday use, such as Mount Maloney Black bowls and jars, contemporary and presumably competing workshops did not obviously differentiate their products through strategies that we might identify as branding, although such differentiation might have allowed one workshop to make claims of superior quality.

Given the potential value of a brand in this context, why did potters not use them? One could speculate that profit motive was too weak in this noncapitalist society or that cultural norms discouraged such differentiation. More convincingly, I would suggest, the structure of the larger economy, particularly in terms of its social dimensions, reduced the utility of branding.

The utilitarian stone tool and ceramic economy of the Mopan Valley seems to have entailed a high degree of intimacy between the consumers and producers. Assuming ceramic marketing paralleled stone tool marketing, consumers obtained both of these types of goods directly from producers, whether the exchange occurred in a workshop or a centrally-located market. This direct connection between consumer and producer is characteristic of a village economy, and it serves to reduce the uncertainties about commodity quality that permeate the bazaar economy and branded economy, in which greater social distance separates consumers from producers. In the village economy, the promise of value that can be achieved through branding derives instead from consumers' relationships with various producers and their knowledge of production processes. The same may have been true of some fancier pottery, like locally produced cream- and orange-slipped painted serving vessels.

Although no formal distance-decay studies have been conducted, it seems likely that utilitarian pottery workshops in the Mopan Valley were fairly closely spaced, and that these potters did not travel to distant markets to sell their wares. This is not particularly surprising, given that pottery vessels are heavy and expensive to transport and that Classic Maya transportation technologies lacked draft animals and wheeled vehicles. Thus, marketplaces were likely dominated by local potting workshops. Another factor could also have advantaged local producers: Guarantees of quality from producers who were also members of the local community plausibly would have inspired more confidence among consumers.

If that was the case, the presence of the occasional nonlocal utilitarian vessel in domestic contexts in the Mopan Valley could reflect either consumers directly obtaining items in more distant marketplaces or maintaining exchange relationships with people in those market zones. In the case of Chaa Creek, the high frequencies of both black- and red-slipped utilitarian pottery suggest that the community's residents frequented different market centres, likely those of Cahal Pech and Xunantunich, which were roughly equidistant from Chaa Creek.

Thus, the production and exchange of utilitarian pottery and chipped stone tools show few structural similarities with either modern or ancient processes of the production of branded goods and brand management. Although the use of Mount Maloney Black vessels may very well have been important in embodying certain aspects of identity (LeCount 2010a), I see little evidence of the more conscious strategies of consumer identification associated with brand consumption (compare Connell 2010).

A possible exception to this pattern is provided by the elaborate polychrome vessels produced in palace schools like those at Naranjo and Buenavista. In both their production and visual differentiation, these vessels recall modern branded commodities. Vessels of a given type were apparently produced in a single workshop, and they required special knowledge and raw materials to make. These facts allowed for greater control over the end-products of each workshop, recalling the emphasis on consistency that is one of the keys to brand management. Meanwhile, the same visual characteristics that allow art historians to identify different palace schools provided Classic-period Maya with a way to know where a vessel was made, analogous in some respects to a brand logo (also Bennet 2008). Finally, these workshops produced serving vessels that were used in public events like feasts and ceremonies. In these contexts, people used the vessels to make statements about identity, class, and political affiliation, much as people today use branded goods to make public statements about their personae. Despite their ostensible similarities with branded commodities, however, these elaborate ceramic vessels do not fit comfortably within our standard definitions of commodities. They were very restricted in their distribution; it is unclear that multiple producers competed in a given marketplace; and the scholarly consensus holds that they circulated largely through gifting networks rather than marketplace exchange.

Conversely, the pottery that least conforms to our notions of a branded good—utilitarian vessels like Mount Maloney Black bowls and jars—was the most commodified, as consumers in a given community acquired nearly identical vessels from multiple workshops. Furthermore, their distribution suggests that consumer choice was strongly conditioned by proximity to the marketplaces where vessels were available. Turning to some of the larger questions taken up in this volume, these data suggest that commodification and branding are independent processes that need not work in tandem. When they do, they can be powerful, resulting in the mass-produced and widely consumed goods that can shape economic decisions, social relationships, identity, and cultural reproduction in important ways (Holt 2004).

The above discussion pertains to locally made items, but some of the goods used in the Mopan Valley were imported. Although most of the valley's pottery and chert stone tools were produced in local workshops, consumers also acquired items that entailed much longer commodity chains (*sensu* Collins 2006). These included goods like obsidian from the distant Guatemalan highlands that we know was exchanged in marketplaces (Cap 2008), as well as goods that probably were exchanged in marketplaces, such as salt and fish from the Caribbean and honey and *cacao* from zones to the east and north. With nonlocal goods like these, production processes were effectively invisible to consumers, and consumer knowledge was heavily mediated by merchants and vendors. Some of these commodities came from multiple points of origin, which raises the possibility, as yet unverified, that branding and/or appellation were important factors in defining these goods and shaping consumer choices.

Why, then, is there not greater evidence of commodities or brands in the Classic-period Mopan Valley, in contrast to the Old World cases discussed by Wengrow and Bevan (this volume)? Although there is ample evidence of long-distance trade of exotic materials such as jadeite, obsidian, salt, and marine shell, Mayanists have generally assumed that bulk goods and staples were not traded regularly over long distances during the Classic period, due to the nature of Maya transportation technologies. Goods had to be carried by human bearers on their backs using tumplines, as the Maya did not use draft animals, wheeled vehicles, or sail boats (Drennan 1984). Thus, for the great majority of the goods exchanged in the Mopan Valley, whether at marketplaces or other venues, consumers likely obtained their wares directly from the producers. In a village economy like this, marked by relatively little social distance between producer and consumer, the uncertainties about quality and pricing that mark a bazaar economy and branded economy are significantly reduced.

This reevaluation of the Maya economy in light of constructs of commodity and brand also leads to some observations regarding the dual model of the Maya economy, particularly the distinction it makes between prestige goods and utilitarian goods. This dichotomy holds relatively well for production: The production of some goods seems to be controlled by elites; most utilitarian goods appear to have been produced in relatively autonomous workshops. Beyond that, however, this distinction begins to blur very quickly: Gift networks likely circulated goods made in attached workshops as well as items collected as tribute and items acquired in the marketplace. Indeed, why should we expect the elite to restrict their gifting to high-value items made in attached workshops? Similarly, it is likely that many so-called prestige items were also available for purchase at markets (also Pyburn 2008 and Sheets 2000 for related critiques).

Finally, beyond its relevance to the topic of commodity branding, our growing understanding of the nature of production and exchange of various types of pottery and stone tools in the Mopan Valley shows the limitations of some traditional archaeological approaches to artefacts. The case study here undermines the distinctions archaeologists sometimes draw between "utilitarian goods" and "prestige goods," as the very same items might be more commodity-like or more gift-like depending on how they were exchanged. These totalising labels also obscure the different meanings a single object can have and the array of social relationships that an object can mediate. The meaning, use, and social efficacy of a given artefact are not inherent to the artefact itself, but rather emerge at the intersection of that artefact's materiality and the social relationships that structure its production, circulation, and use.

Note

1. I would like to thank Andrew Bevan and David Wengrow for their invitation to participate in the symposium that was the impetus for this volume. The data presented here derive largely from the Xunantunich Archaeological Project, directed by Wendy Ashmore and Richard Leventhal, and the Mopan—Macal Triangle Project, directed by Joseph Ball and Jennifer Taschek. I thank the directors and members of both projects for producing such a fine data set and making it accessible through publications and informal conversations. My thinking on the topics addressed in this paper have benefitted greatly from my conversations with Jim Aimers, Andrew Bevan, Bernadette Cap, John Bennet, Christina Dykstra, Douglas Holt, Lisa LeCount, David Wengrow, and Richard Wilk. I would particularly like to thank Anne Pyburn for

encouraging me to think more deeply about consumption as a key force in ancient Maya social processes and for our stimulating exchanges on that topic. Andrew Bevan, David Wengrow, and two anonymous reviewers offered suggestions that improved this chapter considerably. I accept responsibility for any errors of fact and interpretation.

REFERENCES

Abramiuk, M. and W. Meurer. 2006. A preliminary geoarchaeological investigation of ground stone tools in and around the Maya Mountains, Toledo District, Belize. *Latin American Antiquity* 17:335–54.

Ashmore, W. 1995. Settlement archaeology at Xunantunich, 1995. In R. M. Leventhal and W. Ashmore (Eds.), *Xunantunich Archaeological Project, 1995 Field Report*: 10–25. Report on file with the Belize Institute of Archaeology, Belmopan.

Ashmore, W. 2010. Antecedents, allies, antagonists: Xunantunich and its neighbors. In L. J. LeCount and J. Yaeger (Eds.), *Classic Maya Provincial Politics: Xunantunich and Its Hinterlands*: 46–64. Tucson: University of Arizona Press.

Ashmore, W., J. Yaeger, and C. Robin. 2004. Commoner sense: Late and Terminal Classic social strategies in the Xunantunich area. In A. A. Demarest, P. M. Rice, and D. S. Rice (Eds.), *The Terminal Classic in the Maya Lowlands: Collapse, Transition, and Transformation*: 302–23. Boulder: University Press of Colorado.

Ball, J. W. 1993a. Pottery, palaces, and polities: Some socioeconomic and political implications of Late Classic Maya ceramic industries. In J. A. Sabloff and J. Henderson (Eds.), *Lowland Maya Civilization in the Eight Century AD*: 243–72. Washington, DC: Dumbarton Oaks.

Ball, J. W. 1993b. *Cahal Pech, The Ancient Maya, and Modern Belize: The Story of an Archaeological Park.* Ninth University Research Lecture, San Diego State University. San Diego, CA: San Diego State University Press.

Ball, J. W. and J. T. Taschek. 2004. Buenavista del Cayo: A short outline of occupational and cultural history at an upper Belize valley regal-ritual center. In J. F. Garber (Ed.), *The Ancient Maya of the Belize Valley: Half a Century of Archaeological Research*: 149–67. Gainesville: University Press of Florida.

Beals, R. L. 1975. *The Peasant Marketing System of Oaxaca, Mexico.* Berkeley: University of California Press.

Bennet, J. 2008. Palace^{TM}: Speculations on palatial production in Mycenaean Greece with (some) reference to glass. In C. Jackson and E. E. Wager (Eds.), *Vitreous Material in the Late Bronze Age Aegean: A Window to the East Mediterranean World*: 151–72. Sheffield Studies in Aegean Archaeology 8. Oxford: Oxbow.

Cap, B. 2007. Investigations in Buenavista del Cayo's East Plaza. In J. Yaeger (Ed.), *The Mopan Valley Archaeological Project: Results of the 2007 Season*: 44–60. Report on file with the Institute of Archaeology, Belize.

Cap, B. 2008. Investigations in Buenavista del Cayo's East Plaza. In J. Yaeger (Ed.), *The Mopan Valley Archaeological Project: Results of the 2008 Season*. Report on file with the Institute of Archaeology, Belize.

Collins, J. L. 2006. New directions in commodity chain analysis of global development processes. *Research in Rural Sociology and Development* 11:3–17.

Connell, S. V. 2003. Making sense of variability among minor centers: The ancient Maya of Chaa Creek, Belize. In G. Iannone and S. V. Connell (Eds.), *Perspectives on Ancient Maya Rural Complexity*: 27–41. Los Angeles: Cotsen Institute of Archaeology Press, University of California.

Connell, S. V. 2010. A community to be counted: Chaa Creek and the emerging Xunantunich polity. In L. J. LeCount and J. Yaeger (Eds.), *Classic Maya Provincial Politics: Xunantunich and Its Hinterlands*: 295–314. Tucson: University of Arizona Press.

Dahlin, B. H. 2003. Chunchucmil: A complex economy in NW Yucatan. *Mexicon* 25:129–38.

Dahlin, B. H. and T. Ardren. 2002. Modes of exchange and regional patterns: Chunchucmil, Yucatan. In M. A. Masson and D. A. Freidel (Eds.), *Ancient Maya Political Economies*: 249–84. Walnut Creek, CA: AltaMira.

Demarest, A. A. 2004. *Ancient Maya: The Rise and Fall of a Rainforest Civilization*. Cambridge: Cambridge University Press.

Dilley, R. 2005. The visibility and invisibility of production: In different social contexts among Senegalese craftsmen. In W. van Binsbergen and P. Geschiere (Eds.), *Commodification: Things, Agency and Identities: The Social Life of Things Revisited*: 227–42. Berlin/Munster: LIT Verlag.

Drennan, R. D. 1984. Long-distance movement of goods in the Mesoamerican Formative and Classic. *American Antiquity* 49:27–43.

Drennan, R. D. 1988. Household location and compact versus dispersed settlement in prehispanic Mesoamerica. In R. R. Wilk and W. Ashmore (Eds.), *Household and Community in the Mesoamerican Past*: 273–93. Albuquerque: University of New Mexico Press.

Fanselow, F. 1990. The bazaar economy, or how bizarre is the bazaar really? *Man* 25:250–65.

Foias, A. E. 2002. At the crossroads: The economic basis of political power in the Petexbatun region. In M. A. Masson and D. A. Freidel (Eds.), *Ancient Maya Political Economies*: 223–48. Walnut Creek, CA: AltaMira.

Ford, A. and S. L. Fedick. 1992. Prehistoric Maya settlement patterns in the upper Belize River area: Initial results of the Belize River Archaeological Settlement Survey. *Journal of Field Archaeology* 19:35–49.

Garcia, K. L. 2008. Maya pottery specialization and standardization: Using Late and Terminal Classic ceramics from the upper Belize River valley to evaluate market exchange. Masters thesis, University of Alabama.

Gifford, J. C. 1976. *Prehistoric Pottery Analysis and the Ceramics of Barton Ramie in the Belize Valley*. Memoirs of the Peabody Museum of Archaeology and Ethnology, vol. 18. Cambridge, MA: Harvard University.

Graham, E. 1987. Resource diversity in Belize and its implications for models of lowland trade. *American Antiquity* 52:753–67.

Hall, G. D., W. J. Hurst, D. Stuart, R. E. W. Adams, and S. M. Tarka, Jr. 1990. Cacao residues in ancient Maya vessels from Rio Azul, Guatemala. *American Antiquity* 55:138–43.

Hearth, N. 2008. Flaking away at the Mul Chichem household: A preliminary examination of lithic tool manufacture and village-level economy. Paper presented at the 73rd Annual Meeting of the Society for American Archaeology, Vancouver, BC.

Heindel, T. 2008. Investigation of a lithic deposit in Buenavista del Cayo's East Plaza. In J. Yaeger (Ed.), *The Mopan Valley Archaeological Project: Results of the 2008 Season*. Report on file with the Institute of Archaeology, Belize.

Helmke, C. 2001. The last supper: Competitive feasting and the Terminal Classic moulded-carved ceramic tradition in the Central Maya Lowlands. Masters thesis, Institute of Archaeology, University of London.

Helmke, C. and J. Awe. 2008. Organización territorial de los antiguos mayas de Belice central: Confluencia de datos argueológicos y epigráficos. *Mayab* 20:65–91.

Hirth, K. 2009. Craft production, household diversification, and domestic economy in prehispanic Mesoamerica. In K. Hirth (Ed.), *Housework: Craft Production and Domestic Economy in Ancient Mesoamerica*: 13–32. Archaeological Papers of the American Anthropological Association, Number 19.

Holt, D. 2004. *How Brands Become Icons: The Principles of Cultural Branding*. Boston: Harvard Business School Press.

Houston, S. D., D. Stuart, and K. A. Taube. 1989. Folk classification of Classic Maya pottery. *American Anthropologist* 91:720–26.

Houston, S. D., D. Stuart, and K. A. Taube. 1992. Image and text on the "Jauncy Vase." In J. Kerr (Ed.), *The Maya Vase Book*, vol. 3: 499–508. New York: Kerr Associates.

Inomata, T. 2001. The power and ideology of artistic creation: Elite craft specialists in Classic Maya society. *Current Anthropology* 42:321–49.

Jones, C. 1996. *Excavations on the East Plaza of Tikal*. Tikal Report No. 16, Vol. 1. Philadelphia: The University of Pennsylvania Museum.

Keller, A. H. 2006. *Roads to the Center: The Design, Use and Meaning of the Roads of Xunantunich, Belize*. Ph.D. dissertation, University of Pennsylvania. Ann Arbor: UMI.

Kopytoff, I. 1986. The cultural biography of things. In A. Appadurai (Ed.), *The Social Life of Things: Commodities in Cultural Perspective*: 64–91. Cambridge: Cambridge University Press.

LeCount, L. J. 1996. *Pottery and Power: Feasting, Gifting, and Displaying Wealth among the Late and Terminal Classic Lowland Maya*. Ph.D. dissertation, University of California, Los Angeles. Ann Arbor: UMI.

LeCount, L. J. 1999. Polychrome pottery and political strategies in Late and Terminal Classic Lowland Maya society. *Latin American Antiquity* 10:239–58.

LeCount, L. J. 2001. Like water for chocolate: Feasting and political ritual among the Late Classic Maya of Xunantunich, Belize. *American Anthropologist* 103:935–53.

LeCount, L. J. 2010a. Ka'kaw pots and common containers: Creating histories and collective memories among the Classic Maya of Xunantunich. *Ancient Mesoamerica* 21.

LeCount, L. J. 2010b. Mount Maloney people? Domestic pots, everyday practice, and the social formation of the Xunantunich polity. In L. J. LeCount and J. Yaeger (Eds.), *Classic Maya Provincial Politics: Xunantunich and Its Hinterlands*: 209–30. Tucson: University of Arizona Press.

LeCount, L. J. and J. Yaeger. 2010a. A brief description of Xunantunich. In L. J. LeCount and J. Yaeger (Eds.), *Classic Maya Provincial Politics: Xunantunich and Its Hinterlands*: 67–78. Tucson: University of Arizona Press.

LeCount, L. J. and J. Yaeger. 2010b. Conclusions: Placing Xunantunich and its hinterland settlements in perspective. In L. J. LeCount and J. Yaeger (Eds.), *Classic Maya Provincial Politics: Xunantunich and Its Hinterlands*: 337–69. Tucson: University of Arizona Press.

LeCount, L. J., J. Yaeger, R. Leventhal, and W. Ashmore. 2002. Dating the rise and fall of Xunantunich, Belize: A Late and Terminal Classic lowland Maya regional center. *Ancient Mesoamerica* 13:41–63.

Lentz, D. L., J. Yaeger, C. Robin, and W. Ashmore. 2005. Pine, prestige and politics of the Late Classic Maya at Xunantunich, Belize. *Antiquity* 79:573–85.

Leventhal, R. and W. Ashmore. 2004. Xunantunich in a Belize Valley context. In J. F. Garber (Ed.), *The Ancient Maya of the Belize Valley: Half a Century of Archaeological Research*: 168–79. Gainesville: University Press of Florida.

Martin, S. and N. Grube. 2008. *Chronicle of the Maya kings and queens: Deciphering the dynasties of the Ancient Maya*, 2nd edition. London: Thames and Hudson.

McKillop, H. 2002. *Salt, White Gold of the Ancient Maya*. Gainesville: University Press of Florida.

Neff, L. T. 2008. *A Study of Agricultural Intensification: Ancient Maya Agricultural Terracing in the Xunantunich Hinterland, Belize, Central America*. Ph.D. dissertation, University of Pennsylvania. Ann Arbor: UMI.

Neff, L. T. 2010. Population, intensive agriculture, labor value, and elite-commoner political power relations in the Xunantunich hinterlands. In L. J. LeCount and J. Yaeger (Eds.), *Classic Maya Provincial Politics: Xunantunich and Its Hinterlands*: 250–71. Tucson: University of Arizona Press.

Preziosi, A. M. 2003. *Patterns of Style, the Ceramic Record, and Community Differentiation: A Study from Xunantunich, Belize*. Ph.D. dissertation, University of California, Los Angeles. Ann Arbor: UMI.

Pyburn, K. A. 2008. Pomp and circumstance before Belize: Ancient Maya commerce and the New River conurbation. In J. Marcus and J. A. Sabloff (Eds.), *The Ancient City: New Perspectives on Urbanism in the Old and New Worlds*: 247–72. Santa Fe, NM: School of Advanced Research Press.

Reents-Budet, D. 1991. The "Holmul Dancer" theme in Maya art. In V. M. Fields (Ed.), *Sixth Palenque Round Table, 1986*: 217–22. Norman: University of Oklahoma Press.

Reents-Budet, D. 1994. *Painting the Maya Universe: Royal Ceramics of the Classic Period*. Durham, NC: Duke University Press.

Reents-Budet, D. 2000. Feasting among the Classic Maya: Evidence from pictorial ceramics. In J. Kerr (Ed.), *The Maya Vase Book*, vol. 6: 1022–38. New York: Kerr Associates.

Reents-Budet, D. 2006. The social context of kakaw drinking among the ancient Maya. In C. McNeil (Ed.), *Chocolate in the Americas: A Cultural History of Cacao*: 202–23. Gainesville: University Press of Florida.

Reents-Budet, D. and R. Bishop. 2003. What can we learn from a Maya vase? *Archaeology* 56:26–29.

Reents-Budet, D., R. L. Bishop, J. T. Taschek, and J. W. Ball. 2000. Out of the palace dumps: Ceramic production and use at Buenavista del Cayo. *Ancient Mesoamerica* 11:99–121.

Reith, J. C. 2003. SDEP-1, Buenavista del Cayo, Belize: Functional significance and socioeconomic implications of an unusual deposit of chipped-stone microliths from a Classic Maya center in central western Belize. Masters thesis, Department of Anthropology, San Diego State University, San Diego, California.

Robin, C. 1999. *Toward an Archaeology of Everyday Life: Maya Farmers of Chan Nòohol and Dos Chombitos Cik'in, Belize*. Ph.D. dissertation, University of Pennsylvania. Ann Arbor: UMI.

Robin, C., J. Yaeger, and W. Ashmore. 2010. Living in the Hinterlands of a Provincial Polity. In L. J. LeCount and J. Yaeger (Eds.), *Classic Maya Provincial Politics: Xunantunich and Its Hinterlands:* 315–33. Tucson: University of Arizona Press.

Sabloff, J. A. and G. Tourtellot, III. 1992. Beyond temples and palaces: Recent settlement pattern research at the ancient Maya city of Sayil (1983–1985). In E. C. Danien and R. J. Sharer (Eds.), *New Theories on the Ancient Maya*: 155–60. University Museum Monograph 77. Philadelphia: The University Museum of Pennsylvania.

Sanders, W. T. 1989. Household, lineage, and state in eighth-century Copan, Honduras. In D. Webster (Ed.), *The House of the Bacabs, Copán, Honduras*: 89–105. Washington, DC: Dumbarton Oaks.

Sanders, W. T. and B. Price. 1968. *Mesoamerica: The Evolution of a Civilization*. New York: Random House.

Sharer, R. J. 2004. *The Ancient Maya*, 6th edition. Stanford, CA: Stanford University Press.

Sharer, R. J. and C. W. Golden. 2004. Kingship and polity: Conceptualizing the Maya body politic. In C. W. Golden and G. Borgstede (Eds.), *Continuities and Changes in Maya Archaeology: Perspectives at the Millennium*: 23–51. New York: Routledge.

Sheets, P. 2000. Provisioning the Ceren household: The vertical economy, village economy, and household economy in the southeastern Maya periphery. *Ancient Mesoamerica* 11:217–30.
Stuart, D. 2006. The language of chocolate: References to cacao on Classic Maya drinking vessels. In C. McNeill (Ed.), *Chocolate in Mesoamerica: A Cultural History of Cacao*: 184–201. Gainesville: University Press of Florida.
Taschek, J. T. and J. W. Ball. 1992. Lord Smoke-Squirrel's cacao cup: The archaeological context and sociohistorical significance of the Buenavista "Jauncy Vase." In J. Kerr (Ed.), *The Maya Vase Book*, vol. 3: 490–97. New York: Kerr Associates.
Taschek, J. T. and J. W. Ball. 1999. Las Ruinas de Arenal: Preliminary report on a subregional major center in the western Belize Valley (1991–1992 excavations). *Ancient Mesoamerica* 10:215–35.
Taschek, J. T. and J. W. Ball. 2003. Nohoch Ek revisited: The minor center as manor. *Latin American Antiquity* 14:371–88.
Taschek, J. T. and J. W. Ball. 2004. Buenavista del Cayo, Cahal Pech, and Xunantunich: Three centers, three histories, one central place. In J. F. Garber (Ed.), *The Ancient Maya of the Belize Valley: Half a Century of Archaeological Research*: 191–206. Gainesville: University Press of Florida.
Taube, K. A. 1985. The Classic Maya Maize God: A reappraisal. In V. Fields (Ed.), *Fifth Palenque Round Table, 1983*: 171–82. San Francisco: Pre-Columbian Research Institute.
Thompson, J. E. S. 1942. *Late Ceramic Horizons at Benque Viejo, British Honduras*. Contributions to American Anthropology and History 7, Publication 528. Washington, DC: Carnegie Institution.
Tozzer, A. M. 1941. *Landa's Relación de las Cosas de Yucatán: A Translation*. Papers of the Peabody Museum of American Archaeology and Ethnology, vol. 18. Cambridge, MA: Harvard University.
van Binsbergen, W. 2005. "We are in this for the money": Commodification and the *Sangoma* cult of southern Africa. In W. van Binsbergen and P. Geschiere (Eds.), *Commodification: Things, Agency and Identities: The Social Life of Things Revisited*: 319–48. Berlin/Munster: LIT Verlag.
VandenBosch, J. C. 1999. *Lithic Economy and Household Interdependence among the Late Classic Maya of Belize*. Ph.D. dissertation, University of Pittsburgh, Pittsburgh. Ann Arbor: UMI.
VandenBosch, J. C., L. J. LeCount, and J. Yaeger. 2010. Integration and interdependence: The domestic chipped-stone economy of the Xunantunich polity. In L. J. LeCount and J. Yaeger (Eds.), *Classic Maya Provincial Politics: Xunantunich and Its Hinterlands*: 272–94. Tucson: University of Arizona Press.
Vaughn, K. J. 2004. Households, crafts, and feasting in the ancient Andes: The village context of early Nasca craft consumption. *Latin American Antiquity* 15:61–88.
Webster, D. 1997. City-states of the Maya. In D. Nichols and T. Charlton (Eds.), *The Archaeology of City-States: Cross-Cultural Approaches*: 135–54. Washington, DC: Smithsonian Institution Press.

Webster, D., A. Freter, and N. Gonlin. 2000. *Copan: The Rise and Fall of an Ancient Maya Kingdom*. New York: Harcourt College Publishers.

Wells, E. C. 2006. Recent trends in theorizing prehispanic Mesoamerican economies. *Journal of Archaeological Research* 14:265–312.

Wengrow, D. 2008. Prehistories of commodity branding. *Current Anthropology* 49:7–34.

Wurtzburg, S. 1991. *Sayil: Investigations of Urbanism and Economic Organization at an Ancient Maya City*. Ph.D. dissertation, State University of New York, Albany. Ann Arbor: UMI.

Wyatt, A. 2008. *Gardens on Hills: Ancient Maya Terracing and Agricultural Production at Chan, Belize*. Ph.D. dissertation, University of Illinois-Chicago. Ann Arbor: UMI.

Yaeger, J. 2000a. *Changing Patterns of Social Organization: The Late and Terminal Classic Communities at San Lorenzo, Cayo District, Belize*. Ph.D. dissertation, University of Pennsylvania, Philadelphia. Ann Arbor: UMI.

Yaeger, J. 2000b. The social construction of communities in the Classic Maya countryside: Strategies of affiliation in western Belize. In M. A. Canuto and J. Yaeger (Eds.), *The Archaeology of Communities: A New World Perspective*: 123–42. London: Routledge.

Yaeger, J. 2003. Untangling the ties that bind: The city, the countryside and the nature of Maya urbanism at Xunantunich, Belize. In M. L. Smith (Ed.), *The Social Construction of Ancient Cities*: 121–55. Washington, DC: Smithsonian Institution Press.

Yaeger, J. 2010. Landscapes of the Xunantunich hinterland. In L. J. LeCount and J. Yaeger (Eds.), *Classic Maya Provincial Politics: Xunantunich and Its Hinterlands:* 233–49. Tucson: University of Arizona Press.

Yaeger, J. and C. R. Freiwald. 2009. Complex ecologies: Human and animal responses to ancient landscape change in central Belize. *Research Reports in Belizean Archaeology* 6:83–91. National Institute of Culture and History, Belmopan, Belize.

Yaeger, J. and C. Robin. 2004. Heterogeneous hinterlands: The social and political organization of commoner settlements near Xunantunich, Belize. In J. C. Lohse and F. Valdez, Jr. (Eds.), *Ancient Maya Commoners*: 147–73. Austin: University of Texas Press.

Chapter 8

Lincoln Green and Real Dutch Java Prints

Cloth Selvedges as Brands in International Trade

Gracia Clark

Cloth has been traded for centuries over very long distances, sometimes even between continents. Its high ratio of value to weight or volume and its limited perishability has meant that it has often been one of the most significant trade commodities. Cloth from distant sources tends to gain value from its rare or unique features. Yet the many variations available in quality, quantity, and style of cloth pose definite problems for long-distance trade. Specific pieces of cloth are not necessarily interchangeable; strenuous social effort is required to ensure uniformity within even a single town or factory. Creating steady international demand for specific types of cloth has meant conveying knowledge of their identities and qualities to geographically separated populations, and, in the past, this has often been without the benefits of mass literacy, advertising, or global mass media.

Full commodification requires that the buyer can rely on getting what he or she pays for. Quality grades and quantity measures, at least, should be easy to verify. Crossing multiple frontiers of language and culture poses an information challenge for commercial networks. Reliable brand names act as shorthand for standards of quality and quantity that customers would otherwise find difficult to verify for themselves. These standards may include intangible qualities, such as conformity with the latest fashion, which a purchaser relies on the maker to know or to set. Durable utilitarian clothing may require fabrication techniques unknown or invisible to the consumer.

I have only the vaguest idea why jeans last, for example, though I suppose that good-quality cloth and thread, rivets, and felled seams have something to do with it. But I know, by their reputation, that Levi Strauss jeans "wear like iron." They are not going to be bulletproof or waterproof, but they are going to be strong and durable, whereas generic brands may or may not be. My past experience with Levi's makes their advertisement plausible: if I were stranded on a desert island I would want to be wearing

Levi's. They should also be "real" Levi's, the legendary "409" style, not the flimsy, fashionable ones now made by the same company, and certainly not prefaded or preripped. The 409 name has even taken on a meaning of its own in the United States; it appears as the name of a household cleanser, presumably also strong and long lasting.

Brand labels do not solve all the issues of authenticity by any means. Counterfeit designer clothing is rampant, and authenticating individual items through a documented chain of custody is as complex as any bazaar economy. Even where documentation of origin is legally required, as in the United States to keep out blood diamonds or Cuban cigars, it can still be faked.

Provenance naming rights can be hotly contested, as shown by the World Trade Organization controversy over champagne, camembert, and cheddar.[1] The cloth industry abounds with generic category names once based on provenance: calico from Calcutta, plaid cottons from Madras, and paisley shawls (printed with patterns from Indian embroidery). No one now expects denim cloth to come from the French town of Nimes (hence *de Nimes*). Tweed no longer comes from the River Tweed; in fact, premium Harris Tweed could not do so by definition, since its name reflects its provenance from the Shetlands. Fakery has also been practised with basic quality features of cloth, such as weight or stiffness, through the use of sizing or heavy metal rinses.

Distinctive selvedges, the woven edges of a length of cloth, have turned out to be an effective way to establish a kind of brand identity in the cloth trade. Physically an inseparable part of the cloth, they can be made recognisable to an illiterate or partly literate consumer without compromising the use value of the main expanse of cloth. This chapter compares this use of selvedge patterns by medieval guild weavers in Flanders, in northwestern Europe, to the contemporary use of selvedges to mark origin and quality in cotton prints made for the West African consumer. Cloths made in the same region as Flanders, the Netherlands of northwestern Europe, now carry the highest prestige as "genuine Dutch wax."

Medieval Guild Cloths

In late medieval Europe, the most renowned wool cloth was the *grand draperie* from three Flemish towns: Bruges, Ghent, and Ypres (Pirenne 1909–1932). It was traded all over western and eastern Europe and much of Asia, through commercial networks that linked the Hanseatic League, the Italian ports of Genoa and Venice, the Armenian diaspora, and the Silk Road (Previte-Orton 1953: 548). The proximity of Flanders to the English Channel, together with an established expertise in wool weaving from their own sheep, gave the Flemish a near monopoly on England's wool, then its primary export (Des Marez 1927: 39). Flemish merchants collected the year's raw clip each spring from wool fairs in Yorkshire, Lincolnshire, Berkshire, and the Welsh border counties (Peek 1959: 191). They shipped the woolsacks via London or Dover to Flanders, where it was stored in huge cellars to be processed throughout the winter.

The reputation of Flemish wool cloth was based on this access to English wool, which had the long fibres much preferred before the spinning wheel.[2] This made it ideal for strong worsted fabric sewn into clothing that could be handed down in wills because it lasted so long. Some Cistercian monasteries in northern England bred sheep with especially long-staple wool. More than 100 English abbeys were listed on the wool exchange in Bruges (Jouret 1937).

Each of the three towns had its characteristic cloths of standard colour and size, such as Bruges blue, Ghent bruneta, and Ypres white and green (Herent 1904). These cloths were sold by reputation and name all over the continent. Before these licensed town cloths could become so widely recognised they had to be easily identified. A distinctive selvedge pattern was part of the "copyright" enforced for each chartered cloth. This is the edge that is woven on either side of the loom, usually with a tighter weave and extra threads to keep the edges straight and firm (see Figure 8.1).

The selvedge could easily be inspected by a buyer even when the cloth remained folded, with the whole piece sewn together and sealed with the stamp of the local authority. It could be counterfeited, but it could not easily be altered without obviously damaging the cloth. Merchants and wealthy customers could become familiar with the selvedge patterns and seals of popular cloths where they lived without direct contact with Flanders or further knowledge of their methods of manufacture.

The Flemish *grand draperie*[3] gained its prestige from its huge volume rather than its matchless quality. These sturdy, utilitarian cloths were plain tabby weave and solid colours, without the woven patterns of colour or texture available. They were finished smoothly by fulling, stretching, and clipping, which required strong edges. Italian cloths cost more; they were known for elaborate multicoloured finishes; and were softer. But luxury cloth did not wear as well and had a more limited market. The rare fragments of medieval wool cloth surviving (from City of London warehouse excavations) include examples of plain weave with contrasting selvedge threads, but none traceable to Flanders (Crowfoot et. al. 2002).

City authorities and the powerful cloth guilds controlled every aspect of the production in this prestigious industry. Sources and grades of raw materials, weight of thread, and methods of manufacturing were specified in minute detail in municipal regulations under charters granted by the Counts of Flanders (Des Marez 1927; Jouret 1937). Workshops had

a) b)

Figure 8.1. (a) a decorative selvedge on a piece of 20th-century wool cloth; (b) a more normal selvedge on another 20th-century wool (author's photographs).

to be open to the street to facilitate inspection, and for the same reason it was forbidden to mix different grades of wool or to keep wool and cotton in the same workshop (Herent 1904; SEB 1930 [1842]). Secondary towns were limited to making unstandardised cloth or novelty cloths (Coornaert 1930). Those that competed directly with the three "great towns" faced their military expeditions, which destroyed workshops and burned looms (Herent 1904).

Intense proto-industrial development in this small region strained food supplies and political relationships that ensured access to food and wool imports, but it facilitated exports (Pirenne 1909–1932). The three great towns each had a population of more than 20,000 people during the 12th century, the peak of their dominance of European cloth production. The majority were employed in washing, spinning, weaving, and finishing wool cloth (Vanderkindere 1879). In Ghent and Ypres, weavers alone made up more than one-third of the workers (Fris 1913).

Three Flemish cloth fairs followed each other all summer, so that foreign merchants moved from one to the next, serviced by branches of the major banking firms based in Italy. During the 13th century, the autonomous Duke of Champagne founded a popular set of parallel fairs strategically located between Flanders and Italy (Pinot 1906). The Italians eagerly stopped coming as far as Flanders to buy Flemish cloths for resale, let alone to England itself to buy wool for their own luxury cloths (Des Marez 1927: 46; Previte-Orton 1953: 548). After French national integration extended to Champagne in the 14th century, Champagne's fairs quickly foundered (Pinot 1906).

Flanders also gradually succumbed to French domination, during a series of rebellions in the 13th and 14th centuries, after which Flemish refugees were welcomed to England by royal sponsorship (Lipson 1921). Their example inspired a series of named cloths in English towns in the wool-producing regions, where they were deliberately seeded (Doehaerd 1946). The famous Lincoln Green was one of the earliest, a high-quality cloth that was dyed in the wool,[4] and its standards were likewise backed by civic authority and royal charter. Robin Hood and his men in Sherwood Forest may have chosen Lincoln Green for camouflage, but they also supported one of the earliest "buy British" campaigns.

West African Java Prints

The jump from the medieval wool trade to West African cotton prints does not cross as gaping a chasm as one might imagine from the lapse of 500 years. Flemish cloth lost its European dominance in the 14th century, when the spinning wheel made Spanish merino wool more desirable. Bruges citizens turned to making lace and tapestries, and neighbouring Netherlands cities gained commercial dominance. By the end of the 15th century, trade with West Africa had started to fuel European economic growth. Portuguese ships were sailing regularly to Elmina Castle, on the coast of present-day Ghana and providing Columbus with his training voyage. Continuing to push on towards present-day India and Indonesia, Portuguese traders sought to buy pepper and gold at Elmina, and eventually slaves. As they scrambled to find goods that appealed to West African consumers, they began bringing cotton cloth back from India and Indonesia (Harrop 1964).

Cloth was a successful trade item for several reasons. Inland from the coast, West Africans had already been using cloth as currency for centuries in the Sahelian trade networks stretching through North Africa, dating from the era of Carthage. Cotton from West African savannah farms was woven into narrow strips traded in standard long rolls, with some wild silk as a luxury item (Hopkins 1973). Farther south, on the Atlantic Coast, the earliest Portuguese explorers reported that only chiefs wore cloth, whilst commoners then wore finely pounded bark cloth, like Polynesian *tapas* (De Marees 1985). Chinese silks were also bought in quantity but mysteriously disappeared, unravelled to weave into the famous Asante kente cloth and used for underwear (Johnson 1966).

Dutch colonial authorities took over Elmina from the Portuguese in the 18th century and conquered Java in Indonesia by 1830. Whilst ruling both, they recruited Fanti men from the coast around Elmina into the Dutch East India Army as free volunteers from 1855 to 1872 (Kroese 1976).[5] This visible minority of black Dutchmen won medals for their own units, with lively family quarters in Java. Some retired there, but many eventually returned to Elmina, where their old neighbourhood still bears the name Java Hill. They retained the habit of wearing Javanese sarongs, and the fashion spread among their neighbours, sparking imports of Javanese wax prints by Dutch traders.

Intricate Javanese batiks are hand printed using carved wood blocks repeatedly to apply a wax resist pattern for each colour before the cloth was dyed. The multistage process requires a stronger cloth to withstand it, but produces intense colours that last longer than patterns printed in coloured ink, which fade or rub off faster. In factory production, large rollers print the resist pattern onto continuous rolls of cloth in wax or resin. This protective coating still must be removed after dyeing each colour to reveal the complete pattern. Factory-made wax prints remain more colourfast than "Java prints," which show similar patterns but use screen or offset printing to put each colour on.

As the Industrial Revolution progressed, factory-made cloth became significantly cheaper and more profitable than handmade cloth from the colonies. The first imitation batiks were produced in the mid-19th century by Haarlem manufacturer Previnaire, of the Haarlem Printing Company (Kroese 1976). The Glasgow textile trader who began importing them to the Gold Coast towards the end of the 19th century, E. B. Fleming, traced his family business to ancestors from 16th-century Flanders. The British had taken over Elmina in 1872, with the last Dutch holdings in the Gold Coast. Scottish and English textile manufacturers were already imitating Indian cotton textiles like calicos and madras in large quantities (Hopkins 1973). Fleming used Haarlem designs to commission Javanese-style wax prints in Glasgow, and Manchester soon followed. Today, English wax is rated only slightly lower than Dutch wax in Ghana's cloth markets, followed by prints from Ivory Coast, Senegal, Nigeria, Togo, and Ghana.

Buyers in West Africa examine selvedges carefully for evidence of quality categories defined in terms of provenance and method of manufacture, usually not having direct experience of either. African-print selvedge marks use printed words instead of woven patterns, but even illiterate or semiliterate buyers are fully familiar with the quality categories they represent and the price of each. Those same African categories also have a hotly contested history filled with imitations and misnomers. This history created them, ranked them, and adapted them to changing international balances of power.

The display of selvedges reflects a persistent ranking of European over other provenances that goes beyond simple quality assurance. Selvedges that read "genuine Dutch wax" or "real Java print" are frequently worked into the dress design somehow, to flaunt the authenticity of the cloth to those who might not notice. The large printed gold seal marking the end section of the piece may also be displayed prominently in the final product. For those who consider such brand placement rather tacky, it can be made available more subtly in hidden locations like seam allowances or

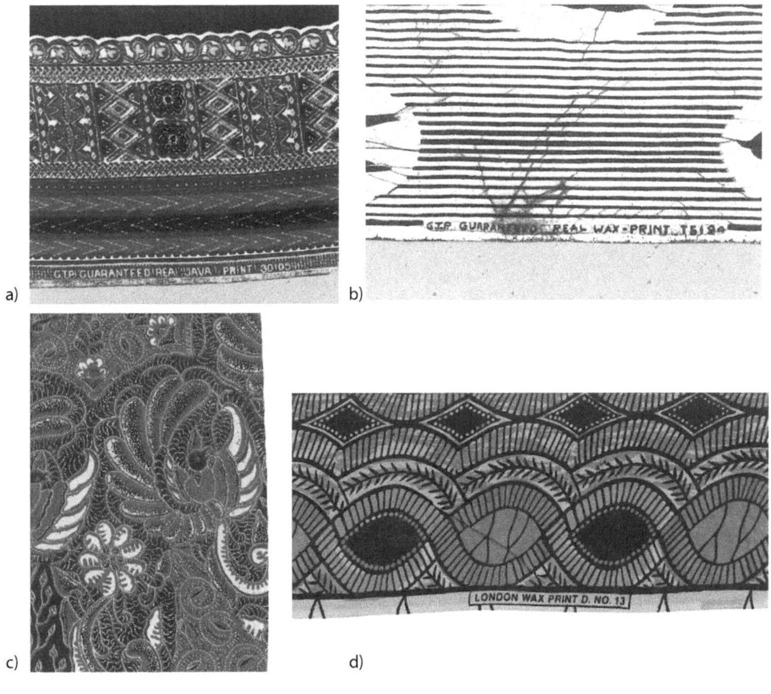

Figure 8.2. (a) a "Real Java Print" from Ghana Textile Producers, bought 1978; (b) a "Real Wax Print" with one colour, from GTP, bought 1979; (c) a sample of Ivory Coast wax print cloth, without selvedge, bought 1979 and then called "Abidjan Wax" in Kumasi; and (d) lighter weight wax print from England, bought 1995 (author's photographs).

shoulder pads. Figures 8.2–8.3 show the selvedge labels and quality differences in African prints, from a blurry local product to the real thing.

The discerning African consumer can reportedly tell the exact quality level of a given sewn outfit from across the room. Quality criteria include weight, stiffness, number of colours used, and the precision of colour placement in the design, called registration. Shoppers can confirm which process has been used by examining the types of imperfections appearing at the selvedges, where the pattern blocks overlap. A wax print also shows distinctive hairline cracks in its large solid colour areas, made by handling the stiff waxy cloth during the dyeing stage. Some printed-on designs include fake cracks, which suggest unsolved technical issues such as colourfastness. Wax prints remain more expensive than the same design applied directly onto the cloth, even by the same manufacturer.

There are several other technical continuities between the trades in Flemish and Javanese cloths besides the use of selvedges. The sale of named cloths by the whole piece was the origin of the English word "wholesaler," compared to the "re-tailleur" or retailer who was willing to cut into the piece and sell shorter lengths. The "whole piece" and "half piece" are still common units of sale in the West African cloth trade today, and a whole piece is still sewn together and sealed by the manufacturer.

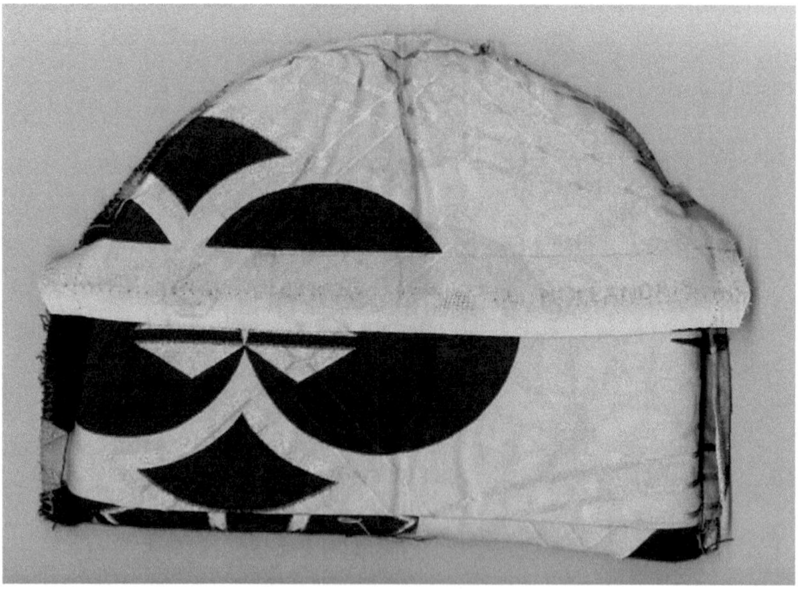

Figure 8.3. Real Dutch wax from Raynes of Holland, bought 2001 (photograph courtesy of Beverly Stoeltje).

Today's 12-yard "whole piece" is much more than needed for one "traditional" West African three-piece outfit. Interestingly, 12 yards is also the standard length of an Indian sari. Two friends can express their closeness by buying a whole piece and splitting it to wear together to social events. Originally, Ghanaians wore their cloth in large squarish pieces, sewing together several lengths, like they did their narrow strip cloth, and this is still the most formal attire for ritual occasions.

Naming Traditions

The relationship between original, traditional, and genuine has been inverted and reinverted in the history of the "African print." European manufactured cloth has completely displaced the original Javanese batiks by now, so that "our traditional African dress" is at its most authentic in Ghana with "genuine Dutch Java print" cloth. This cloth is not traditional European cloth either, but made specifically for African tastes. There are markedly different patterns for the West African and East African markets, and some that imitate West African handmade cloth—kente, adinkra, bogolan, and others.

Ironically, Indonesia now makes cheap imitations of these "African prints" in lighter weight cottons sold by the yard from large bolts. Indonesia or other Asian countries also imitate cheaply traditional European cloth types popular with West Africans. Damasks, brocades, and eyelet lace have been reinterpreted to mean affluence and power for African men as well as women, even when woven in Singapore. They have their own quality standards but less firmly bounded categories.

Fabrics sold by the yard, not by the whole piece, are not even called cloth. The English word "materials" is used even for expensive yard goods, rather than "cloth," or the Akan word *ntama*. Real cloth retains its value and was used as a store of value through decades of rampant inflation. It can be stored in trunks, handed down, and resold in time of need, especially if it remains uncut.

The practice of selling in whole pieces supports both the high volume of trade in West African prints and their extensive market penetration. Buying in bulk would be cumbersome indeed if every sealed piece of cloth had to be opened for its dimensions and quality to be verified. Only those with close personal relationships to the supplier could feel confident enough to return defective goods. In this context, product branding does not bring decommodification, but reduces the need for personal economic ties to compensate for risk from indeterminacy.

Retail shopping and bargaining are also greatly simplified because the buyer can compare prices and patterns within a familiar framework of categories that stipulate quality and quantity (see Figure 8.4). Even sellers

Figure 8.4. A cloth stall in Kumasi Central Market (author's photograph).

who sell cloth in 2-, 3-, or 4-yard lengths cut from whole pieces generally keep these shorter pieces in their original folds to verify their length and show the selvedge. Dutch wax is virtually never cut.

One of the more flexible quality issues that affects the price of an African print cloth is its specific pattern. Some patterns are more fashionable than others, and fashions change. Wholesalers try to negotiate exclusive rights to a new pattern they think will be popular when they order it from their distributors.[6] Having the latest pattern brings a premium to the trader and to the wearer, who has it sewn up in the latest style. It sells faster and at a higher price than generic prints of the same quality.

Patterns are themselves subject to marketing pressures and strategies and have important branding aspects (Kroese 1976). A catchy local name for a pattern, linked to current events or proverbs, makes it more attractive. The glamour of naming patterns is also effective within the personalised craft realm of kente cloth, normally woven to order. The customer can choose among patterns named for ancient proverbs or for contemporary events, such as the "Nkrumah loves Fatima" pattern dating from President Nkrumah's marriage.

Other patterns are old standards and some become dated. Like outdated styles of sleeve or neckline, they advertise how long it has been since the wearer could afford a new cloth. Still, many Ghanaians fall into that category, and old patterns reappearing may indicate the extent of these hard times as well as the cycle of fashions.

Women have bought good quality cloth as a store of value for many decades. It held its value much better than money during decades of high inflation after 1960, despite losses suffered from moths, mice, and mould. Full pieces with sewing and seals intact sold as new, more or less. Sewing a cloth to fit the wearer and her time does not destroy its commmodity value or require later recommodification.

A vigorous market exists for second-hand cloths, already sewn into styles of long ago or not so long ago. These can sometimes be modified to look more up-to-date, but usually betray their second-hand origin. Wearing a good older cloth still carries more prestige than the same type of clothing sewn from cheaper fabrics. The inexpensive local tie-and-dye cottons have a nickname "My husband's work is spoiled," implying he could not afford to buy a real cloth. Second-hand sewn cloths are also more respectable for formal occasions than second-hand clothes of European or American origin.

Discussions of commodification in Europe need to consider the links between the neighbourhood garage sale and the global market in second-hand clothing, cars, and other goods. Second-hand clothes from America outrank those from Europe, except for custom-made British suits. The distribution channels for second-hand clothes are complex, repeatedly crossing boundaries between the formal and informal sectors, capitalism, and the gift (Hansen 2000). The same channels carry and price appropriately cast-offs, donated clothes, brand knockoffs, and unsold clearance items. Point of origin, production process, previous use, and brand labelling all become attributes of a commodified item that is then carefully evaluated in the marketplace by retailers and consumers at each purchase.

Who Controls the Brand?

The branding features of African prints made West Africans loyal customers of European firms, but did not simply subordinate them to colonial designs. Branding also gives African customers some leverage over the distant producers. The reputation of the brand, like that of the supplier, becomes something that can be lost or won. Dutch factory owners might discipline their workers very differently than medieval guild masters, but their global sales still depended on their reputation for delivering a consistent product. Systems of control had to be adequate to maintain standards of quality that the final consumer in West Africa was quite capable of evaluating. Brand magic might be marked by the selvedge sewn and sealed in Holland, but substantial qualities had to correspond: weight, size, and precision of printing, plus the continuous innovation in designs that met consumer demand.

The attachment of West African buyers to branded commodities that assured them of high-quality standards for their unexpectedly sophisticated tastes sometimes frustrated the European firms operating there. Early Dutch traders around 1600 warned that copper pots, one of their staple items, had better be of good solid thickness and not the cheap English kind, or the West Africans would refuse them and the trader would end up bringing them back (De Marees 1985). A political scandal arose in the 1940s when British firms tried to unload unpopular kinds of cloth by conditional sales, whereby customers had to buy a certain amount to obtain some of the scarce cloth they really wanted (NAA 4, 5, and 6). As those same firms contended with postwar shortages, they complained that Gold Coast customers insisted on premium Schwinn bicycles and would not settle for cheaper brands (UAC 1950). That brand loyalty was conditional and did not prevent them from switching to Chinese bicycles in the 1980s when incomes fell.

The African buyer then and now does not passively accept commodities branded for European tastes but pressures manufacturers to continue and elaborate brands that suit them. The biggest wholesalers are consulted by factories about future designs. They are the ones who make or break a new pattern by their factory order, their local names, and their decisions to reserve or ration access. Two fairly recent video documentaries give different perspectives on the powerful role of these leading wholesalers, often called Mama Benz or Nana Benz for their iconic cars.[7]

Patterns of cloth are designed for particular countries or ethnic preferences. In Ghana, distinctive aesthetics and colour palettes signal cloth patterns intended to appeal to Asante or Ewe buyers or northern communities. Identical cloths are ordered in bulk to serve as an emblem of unity for members of occupational or religious groups, or for lineage members or grandchildren at an important funeral. Important national or religious events such as inaugurations, papal visits, and centennials merit a commemorative cloth. The cloth marking the 150th anniversary of the Presbyterian Church in Ghana, for example, featured large medallions with portraits of the founding fathers. It remained conspicuous at Presbyterian Church services for decades afterwards.

Dutch wax remains the gold standard of traditional West African dress long after Dutch commercial dominance has waned in other respects. Ghana's cloth traders may travel by air to London for the best selection and price, but they continue to pay more for Dutch wax than for London wax. Dutch wax is expensive, beyond the means of university lecturers except for their one best cloth. Its everyday use has become the emblem of success for big businesswomen, cabinet ministers' wives, and cocaine traffickers.

Some Dutch manufacturers tried to promote it as a high fashion item but with limited success.[8] The core clientele for Dutch wax is middle-aged and proud of it. They are far more concerned to display their wealth than to look young and thin, which carries little prestige among their peer group. Increasingly, male businessmen are also adopting this type of cloth for styles originally executed in solid-colour cloth marketed as masculine. Younger women who might have considered European-style dresses are tempted with styles that echo their aspirations to combine modernity with their prosperous mothers' status. One neotraditional style from the Vlisco website is modelled in Black Star Square, the shrine of Ghana's independence (Vlisco 2008). It highlights a slim but rounded figure, and its details echo popular up-and-down styles along with traditional waist beads. By combining Western fashion aesthetics with West African aesthetics of display, such representations help keep the brand viable in contemporary elite circles. It still appeals to an elite cohort strongly invested in both Western and African directions, whether materially or in imagination.

More or Less Than a Brand?

As in the first case of medieval Flemish wool cloths, the multilocal origins of "African print" cloths have interlocked firmly to form a stable, unitary identity that still encodes and displays them. In this second case, Dutch wax, the generic Dutch or *Hollandais* provenance identity incorporates a number of corporate brands that are familiar to relevant consumers. This kind of integrated quality category was no blind artefact of the market, but in each case was historically constructed and culturally reinforced with considerable care and attention.

Like contemporary industrial brands, the intangible quality of reputation provided a transcendent element that communicates itself to the consumer of "Dutch wax." The authentic name on the selvedge releases some of these magic qualities, but the concrete material qualities of the cloth laid the foundation. Both Dutch wax and Levi's are vulnerable to the same threats to the value of brand authenticity from counterfeits, replicas, and imitative competitors aiming to appropriate that mystique.

The convenience to consumers of branded merchandise is one factor that greatly extends its market penetration, especially when cultural difference and distance are significant factors. Customers so placed may have little direct access to information on manufacturing techniques, towns of provenance, or standard measures of size and weight. They can become familiar with this kind of category, its internal differentiation and their visible markers within their home environment. By monitoring indicators such as woven or printed selvedges, illiterate consumers around the globe can buy with enough confidence to sustain a vigorous trading network at every level from global to local village.

This relationship undeniably subordinates Africans on one level to manufacturers in more industrialised areas, not least by displacing rural populations and submerging their sources of income. Still, it also gives consumers some leverage. Pressure from consumers to maintain a supply of goods such as cloth, in forms that have meaning and utility in their own lives, does help protect their standard of living as they see it. Without "Dutch wax," an Asante funeral would not be the same. When medieval wars dispersed Flemish weavers into England, the venerable names of their conquered homeland could not compete with the actual skills they took with them. Despite the arguably nefarious intentions of contemporary advertising, effective brands operate in a dialectic process that has both material and ideological parameters.

Notes

1. Various world trade organizations are hearing a series of disputes over the rights of similar products from outside the original provenance areas (champagne for sparkling wine and camembert and cheddar for types of cheese, among others) to use those names as a generic designator for their products.
2. Longer hairs make a stronger thread when the raw wool is combed and then spun with a drop spindle. The spinning wheel produces thread much faster but breaks the longer hairs, removing this advantage. Carding also breaks the hairs but prepares wool for spinning much faster than combing it. The two technological changes meant fewer spinners could keep a weaver in thread, and meant that the shorter, finer, denser wool of Spanish merino sheep competed effectively with English fleeces.
3. This is, literally, great cloth business.
4. This dyeing method used more dye, but the colour did not wear off as the cloth became worn.

5. Earlier Elmina recruits (1837–1843) had mainly been former slaves and so had distant ethnic origins. They could buy their freedom with the recruitment bonus.
6. Both of the videos titled *Mama Benz* show this expectation of exclusive rights.
7. Two excellent videos confusingly carry the same title *Mama Benz*. One subtitled *God Gave Her a Mercedes* (1993) profiles a leading Lomé market wholesaler. The second, subtitled *The Taste of Money* (2002) recounts the unsuccessful attempt to bypass such a wholesaler in Ouagadougou, led by the local manager of a Dutch firm.
8. The video subtitled *The Taste of Money* shows this image disjuncture in telling detail.

References

Coornaert, E. 1930. *Un Centre Industriel d'Autrefois, La Draperie-Sayetterie d'Hondschoote*. Paris: Les Presses Universitaires de France.

Crowfoot, E., F. Pritchard, and K. Staniland. 2002. *Textiles and Clothing, c.1150–c.1450*. Rochester, NY: Boydell Press.

De Marees, P. 1985. *Chronicle of the Gold Coast of Guinea*. Translated by A. Van Dantzig and A. Smith from the 1602 edition. Oxford: Oxford University Press.

Des Marez, G. 1927. *La Droit. Prive a Ypres au XVIIIe Siecle*. Braine-l'Allaud, Belgium: Rene Berger.

Doehaerd, R. 1946. *L'expansion Economique Belge au Moyen Age*. Brussels: La Renaissance du Livre.

Fris, V. 1913. *Histoire de Gand*. Brussels: G. Van Oest.

Hansen, K. 2000. *Salaula*. Chicago: University of Chicago Press.

Harrop, S. 1964. The economy of the West African coast in the 16th century. *Economic Bulletin of Ghana* 8:15–29.

Herent, J. 1904. *La bataille de Mons-en-Pevele*. Lille, France: Imer de Lefedre-Nucrocq.

Hopkins, A. G. 1973. *An Economic History of West Africa*. New York: Columbia University Press.

Johnson, M. 1966. The ounce in 18th century West African trade. *Journal of African History* 7:197–221.

Jouret, G. 1937. *Histoire Economique de la Belgique*. Mons, France: La Province.

Kroese, W. T. 1976. *The Origin of the Wax Block Prints in the Coast of West Africa*. Hengelo, Netherlands: Smit.

Lipson, E. 1921. *The History of the Woollen and Worsted Industries*. London: A. & C. Black.

Mama Benz & the Taste of Money. 2002. New York: Filmmakers Library.

Mama Benz: God Gave Her a Mercedes Benz. 1993. New York: Filmmakers Library.

Peek, G. S. 1959. *The Anglo-Flemish Conflict (1270–1274) and Its Effects*. Ann Arbor, MI: University Microfilms.

Pinot, J. 1906. *Etude Historique sur les Relations Commerciales entre la Flandre et la Republique de Genes au Moyen Age*. Paris: Picard.

Pirenne, H. 1909–1932. *Histoire de Belgique*. Brussels: H. Lamertin.

Previte-Orton, C. W. 1953. *The Shorter Cambridge Medieval History*. Cambridge: Cambridge University Press.

Societé d'emulation de Bruges (SEB). 1930 [1842]. *Collection des Keuren ou Statuts de Tous les Metiers de Bruges*. Gant [Ghent], Belgium: Annoot-Braeckman.

United Africa Company (UAC). 1950. Merchandise Trading in West Africa. *Statistical and Economic Review* 5 and 6.

Vanderkindere, L. 1879. *Le Siecle des Artevelde*. Brussels: J. Janssens.

Vlisco. 2008. The Dutch Wax. http://www.africanews.com/site/list_messages/1560 (accessed 15 March, 2008).

Sources from the Ghana National Archives, Accra, cited under its former system of classification, as found in 1980:

NAA4. No. 0866 ST17. Complaints against profiteering by trading firms. Anonymous letter to the Governor, 23/9/41.

NAA5. No. 0866 ST17. Letter from President of the Joint Provincial Councils to the Secretary for Native Affairs, 20/11/41.

NAA6. No. 0028 SF8. "Irregularities in Import Control." Motion by Hon. Dr. J. B. Danquah, 26/3/47.

Chapter 9

Of Marks, Prints, Pots, and Becherovka

Freemasons' Branding in Early Modern Europe?

Marcos Martinón-Torres[1]

Introduction

Most archaeologists and anthropologists accept that the production and consumption of material culture provide an important mechanism for the negotiation of personal and collective identities and bonds. In making, exchanging, and using things, individuals construct and express their own position with reference to wider social structures. Explicitly or implicitly, commodity consumption is always entangled with cultural constraints as much as with economic networks; thus, in studying consumption, we may obtain insight not only into economic organisation, but also into human aspirations and beliefs, or the development and transmission of ideas in different contexts.

The contributors to this volume explore the concept of branding as a particular mechanism for structuring the marketing of mass-produced, replaceable goods in social contexts where there is often a significant separation between producers and consumers. In these situations, the labelling, packaging, or other marking of substitutable commodities seem effective ways of allowing consumers to navigate the market and identify desirable goods. However, branding is not just a form of consumer protection against adverse selection or fraud; importantly, branded commodities foster loyalty and mutual recognition among consumers, developing bonds that are reinforced by shared values and expectations, whether real or assumed.

Researching the social and intellectual ramifications of commodity branding becomes arguably more difficult when we deal with past societies. This is probably due to the fragmentary nature of the data as much as to our presentist bias and comparative lack of familiarity with how social relationships might structure the consumption of goods in cultural contexts alien to our own. However, this is not an altogether impossible endeavour.

For example, Bevan's study (this volume) convincingly explores how the values assigned to different branded products changed as these goods moved across different social sectors in the Bronze Age Mediterranean, and how those values were contingent on the specific mechanisms structuring broader social relationships.

This chapter is concerned with the development of specific values and cultural ties behind what appears to be an early modern European "brand." In particular, I investigate the institutional structures and changing ideas that may explain the presence of a stamp or monogram found in a range of goods, including books, ceramics, and metalware, that have been consumed across Europe during several centuries since late medieval times. Thus, the first peculiarity of this case study is that it documents a possible branding strategy that cuts across different producers and material categories over an extremely long time span. In this case, the link between the branded goods is clearly not a single producer or the type of product but the ideas associated with them: namely, a complex mix of religious tenets, late alchemical imagery, and emerging freemasonic ideals. The second peculiarity is the claim that, in accordance with much alchemical and freemasonic practice, the ideas expressed by the stamp in question are secretive and understandable only to those initiated in the relevant codes. Based on these and other considerations, this chapter will recognise the value of "branding"as a useful model to explain patterns of commodity production and consumption, but it will also highlight some of the great variability of human behaviours that take place under such single conceptual umbrella.

Background

The history of freemasonry remains elusive. This is partly because of the secrecy inherent to its initial organisation and much of its later history, together with its elaborate language, complex symbolism, and a wealth of foundational myths that confuse the historian. A further complication stems from the fact that most of the traditional works on the history and symbolism of freemasonry were written by masons, naturally more familiar with the topic than most, but sometimes more stirred by laudatory enthusiasm than by academic rigour.

In the last couple of decades, some exemplary work, most notably by Margaret Jacob (1991, 2006) has brought greater clarity to the origins, social implications, political activities and historical development of masonic lodges. One of the aspects that we have begun to understand better is that of the institutional background to their formation. It seems clear that freemasonry relied on the structure of earlier arts and crafts guilds that, from the late 17th century, started to accept selected fee-paying laymen in order to face financial burdens. Progressively, in a changing market economy, the professional specialisation and economic role of the guilds decreased, whilst their role as structures for the socialisation of "gentlemen" developed.

Another subject of historical enquiry has been the links between freemasonry, alchemy, and the hermetic tradition, even though these connections still remain to be explored in detail (Yates 1964, 1975). It is widely accepted that masonic myths and symbols draw substantially from alchemical imagery, as the secretism, austerity, allegorical language, and allegedly high aspirations of both alchemists and freemasons have much in common.

As far as I am aware, however, no work has as yet attempted to investigate the multiple connections between these three dimensions: the late medieval guild, the alchemical laboratory, and the early modern masonic lodge. This network is the particular focus of this chapter. Using archaeological material culture together with historical and iconographic sources, I discuss the transfer of ideas and symbols between alchemists, potters, printers, freemasons, and other individuals, in what appears to be a peculiar case of freemasons' branding.

Even so, the branding of material culture by early modern freemasons is a slightly odd subject of inquiry, and this study may require some form of justification. The original spark behind this chapter was simply a series of initially perplexing coincidences and chance discoveries that attracted my curiosity in the course of a different project, rather than a planned research programme. I report my ideas below in the order they developed, whilst accommodating more extended digressions and acknowledging caveats where relevant.

Crucibles, Guilds, and Stamps

The starting point of this research was a project of archaeological and scientific analyses focused on the archaeology of alchemy and also aimed at reconstructing the production and consumption of crucibles in the post-medieval world (Martinón-Torres 2005, 2007; Martinón-Torres and Rehren 2005, 2009). Crucibles are ceramic vessels normally used for chemical and metallurgical operations. From the late Middle Ages onwards, we can document an increase in the quality and diversity of available crucibles, with relatively large numbers of them recorded archaeologically from alchemical laboratories, metallurgical, and jewellers' wokshops, glass-works and mints, where they were used for mixing, melting, calcining, reducing, evaporating, or otherwise processing substances under high temperatures (Figure 9.1). Two main crucible producers can be identified whose products were exported across early modern Europe and even to colonial America: (1) the famous mullite crucibles produced in the region of Hesse, in central Germany (Martinón-Torres et al. 2006, 2008; Stephan 1995), and (2) the graphitic crucibles manufactured in Bavaria and surrounding regions (Bauer 1983; Martinón-Torres and Rehren 2009). The latter group is of particular relevance to this chapter, given the presence of a stamp imprinted on the exterior base of some examples, which has encouraged the research reported here.

The stamp in question is characteristically organised around the number 4, usually with a second horizontal crossbar, and two initials on either sides of the vertical stem. This design is usually enclosed by a niche-shaped outline, with the longest axis being between 2 and 3 cm in length. In some cases, the stamp is more neatly delineated, as if imprinted with a metallic, rather than a ceramic, die. Variations in this design are noticeable on different crucibles, chiefly in the form of different initials (or the same ones in

Figure 9.1. An assayer's workshop as depicted in Georgius Agricola's 1556 *De Re Metallica* (Hoover and Hoover 1950). Several crucibles are shown on the foreground—one of them upturned, revealing a mark on the base.

slightly different style) as well as via the addition of further elements such as an anchor at the base of the mark (Figure 9.2).

Stamped crucibles of this type are frequently found in and around the village of Obernzell, in Bavaria (southern Germany), where they are known to have been made (Bauer 1983). However, we have further examples in laboratories and workshop remains excavated in Austria, Croatia, Switzerland, and the United Kingdom as well as a remarkable collection excavated at the Imperial Mint of Rio de Janeiro, Brazil (Cotter 1992; Lima and da Silva 2003; Martinón-Torres 2005; Martinón-Torres and Rehren 2009). Of these archaeological examples, some remain undated at present, but those dated with confidence have been ascribed to the 16th century and later.

It is not surprising for these crucibles to bear a maker's stamp. Certain early modern crucible makers merely seem to have followed the custom of many other contemporary potters in marking their products as a part of

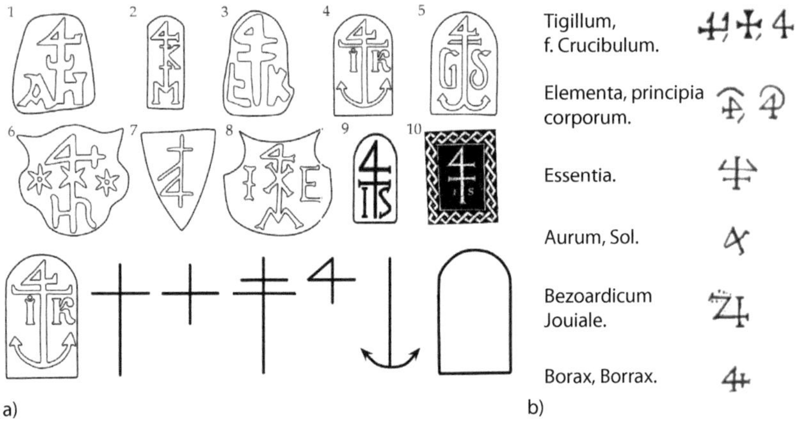

Figure 9.2. Symbols and emblems on early modern material culture. (a) 1–5, several stamps recorded on the bases a variety of crucibles (after Cotter 1992 and Bauer 1983); 6, seal of the Dutch merchant Jacob van Cosveloe (Milne 2003); 7, seal of Hans Hell of Aflenz, Austria; 8, central emblem on a *Bartmannskrug* by Jan Emens Mennicken (Gaimster 1997); 9, symbol imprinted on the base of some of the 18th–19th century crucibles recovered in Rio de Janeiro (after Lima and da Silva 2003); 10, emblem used by John Siberch in Cambridge during the 16th century (Fisher 1521). The bottom row shows the stamp of a crucible from Obernzell (Austria), and the series of symbols it includes: from left to right, the stamp itself (with the initials IK), a Latin cross, a Greek cross, a double cross, a number four, an anchor, and a niche-shaped outline. (b) some examples of alchemical symbols illustrated in the *Medicinisch-Chymisch-und Alchemistisches Oraculum* of 1755.

their guild customs. Gathering particular importance after the European urban and economic revival of the 12th and 13th centuries, craft guilds had operated as self-regulatory corporations that protected both their own trade and their customers by organising apprenticeship, manufacturing standards, and controlling the quantity, quality, and price of their stock as well as offering other services to their members such as protection against professional encroachment or even religious ceremonies for the deceased (Black 1984; Epstein 1991; Gadd and Wallis 2002, 2006). In a market flooded with counterfeit goods and where trade was channelled through intermediate merchants, potters would use their guilds' stamp on their products, among other conspicuous characteristics, as an indication of reputation and quality for potential buyers (Richardson 2008). Merchants (Girling 1962), printers (Roberts 1893), stoneworkers (Jones-Baker 1996), and cloth weavers (Clark, this volume) amongst others, also marked their products.

The use of stamps to mark "authentic" or reputable goods seems particularly appropriate for crucibles, as these were the key instruments of alchemy—an area of activity where serious scholarly work was encumbered by commonly recognised patterns of charlatanry, fraud, and deceit, and also an international market niche of very substantial worth. As a case in point, it is estimated that millions of crucibles were imported from Germany into Britain alone during the post-medieval period, causing such a financial burden that the Royal Society of Arts actively promoted the manufacture of crucibles by local potters—albeit with limited success (Cotter 1992).

It may also be significant that, out of the innumerable metallurgical tools illustrated in the nearly 300 woodcuts in Agricola's *De re metallica*, crucibles seem to be the only instruments clearly showing a stamp (Figure 9.1). The role of the stamp as a propositional device in crucibles thus seemed an established and understandable practice in the context of a large-scale, long-distance, competitive market in standardised goods. Based on such premises, my initial interpretation of the stamp found on the archaeological crucibles was straightforward and even offered some effort-saving advantages to my initial materials science project—those crucibles bearing the familiar stamp would no longer need to be subjected to scientific analyses to determine their provenance, as the stamp itself would prove their Bavarian origin.

Things, however, turned out to be more complicated, when similar symbols began to be noticed in completely different media—often, on objects that had no known connections with Bavaria (Figure 9.2). One of the most startling cases appeared in the mark used by John Siberch, a printer who worked in Cambridge in the early 16th century (e.g., Fisher 1521). Apart from the niche-shaped outline (absent in the printed version), the symbol on the last page of some of his books is virtually identical—even in the initials, IS— to that found on the bases of the crucibles produced over 200 years later and recovered in the mint of Rio de Janeiro, Brazil (Lima and da Silva 2003; Figures 9.2.9 and 9.2.10).

More examples continued to emerge: For instance, one of the emblems on the famous stoneware *Bartmannskrug* or "Bearded Man Jug" manufactured in the late 16th century by the Belgian potter Jan Emens Mennicken (1540–1593), now at the Victoria and Albert Museum in London (Gaimster 1997: 225, 236), again bears a resemblance that seemed too strong to be coincidental (Figure 9.2.8). A similar mark is engraved on the wall of the west tower of the Obernzell castle in Germany, next to an annotation of the year 1573 (Bauer 1983: 36), and depicted in the seal of the Dutch merchant Jacob van Cosveloe (Milne 2003: pt. 3; see Figure 9.2.7), or in that of Hans Hell of Aflenz, Austria (exhibited in Propstei Aflenz in 2006; see Figure 9.2.6), as well as many other seemingly unconnected instances. In short, variations of this peculiar symbol seemed to have been employed by early modern goldsmiths, engravers, sculptors, potters, glassmakers, painters, drawers, stonemasons, printers, and many other specialists.

Deconstructing a Symbol

Researching the possible origins and meanings of the symbol, I was drawn into previous work on the role of marks and monograms in different arts and crafts. Interestingly, others had been puzzled and somewhat frustrated before me. Amongst these, William Roberts (1893) deserves a more extended quote, as he, in turn, shows that that uncertainty over the implications of this charismatic mark is nearly as old as the symbol itself:

> Finally, in his introduction to [1730] *Thesaurus Symbolarum ac Emblematum*, Spoerl asks,
>
> Why are the initials of a printer or bookseller so often ... surmounted by a cross? Why at the extreme top of the cross is the lateral line formed into a sort of triangular four? Why, without this inexplicable sign, has the cross a number of cyphers, two, or even three, cross-bars? Why should the tail of the cypher 4 itself be traversed by one or sometimes two perpendicular bars which themselves would appear to form another cross of another kind?
>
> All these are problems which it would be exceedingly difficult to solve with satisfaction. We do not propose offering any kind of explanation for these singular marks; but it will not be without interest to point out that among the more interesting examples are those used by [printers or booksellers of Paris, Lyon, Strassburg, London,

> St. Albans, Antwerp, Leipzig, Barcelona, Seville] and hundreds of others that might be mentioned. (Roberts 1893: 23–24)

Rather more enlightening was the work by Lima and da Silva (2003). As in my case, their research began with the study of crucibles bearing the same mark, for which they then sought parallels and meanings. To my knowledge, they were the first to suggest connections with alchemy and freemasonry. Some of their statements, particularly with regards to the history of both alchemy and freemasonry, reflect historiographic misconceptions (for recent reviews, see respectively Principe and Newman 2001 and Jacob 2006), and they were not acquainted with everything we now know about the production and trade of crucibles in the early modern world (see Martinón-Torres and Rehren 2009). However, their work remains exemplary in its emphasis on the fact that the mark can be broken down into a series of simpler, superimposed symbols whose individual social and economic lives are worth separate consideration before attempting an overarching interpretation. What follows thus owes much to their approach, although I have moved away from a number of their identifications and interpretations.

As illustrated in one of its most complex versions in Figure 9.2, the stamp can be deconstructed into a series of simpler shapes: Apart from the letters (most likely the initials of the master or sponsor of the work), there is a Latin cross, a Greek cross, a double cross, a number 4, an anchor, and, frequently, a well-delineated niche. Each of them, and their combination, appear to become meaningful in the context of Renaissance and later religion, alchemy, and freemasonry.

The Latin cross (i.e., that with the longer vertical axis) is universally recognised as a representation of Jesus Christ and has been historically employed as the defining emblem of Christianity. The Greek cross (*crux quadratta* or equi-axed) is currently used more frequently by the Eastern Orthodox Church, but was also a common symbol used by early Christians. More generally, the cross is often seen as "the cosmic symbol par excellence" and a point of integration between the rational and the spiritual (Cooper 1990: 45). It is worth noting that in English as in Latin and many other languages, the word for "crucible" seems etymologically rooted in the term "cross," and both of them (the crucible and the cross) are associated with ideas of synthesis and purification in religious as much as alchemical contexts.

The double cross may also convey various meanings, which are not mutually exclusive. When the top crossbar is smaller than the main one, this cross is often termed a "Patriarchal cross," and is used by Eastern Orthodox Churches but also by Roman Catholic archbishops to differentiate their office. Here, the top bar represents the plaque bearing the inscription INRI of the Crucifixion. The double cross, however, may also constitute a deliberate superposition of the Latin and the Greek crosses. If one accepts that the Latin cross represents the microcosm, and the Greek cross represents the macrocosm (Lima and da Silva 2003: 47), then their superposition might be a statement of the theory of identity between both: The idea that the same patterns are reproduced at all levels of the universe was strongly present in Classical Greek philosophy and in subsequent Neoplatonism and it was central to later alchemy, which sought to reproduce in the laboratory the phenomena (such as metal transmutation) that were believed to take place in nature.

Turning to the number 4, this may also combine religious iconography with alchemical tenets. On the one hand, the figure can be seen as a Latin cross to which a diagonal bar has been added, thus subtly concealing the religious icon. Such use of the number 4 as a way of sharing religious values in disguise is thought to have been employed by 16th-century printers in Lyon during the time of the Reformation (Lima and da Silva 2003: 20), and it is plausible that more people could have adopted this symbol in contexts of religious struggle. On the other hand, 4 is a totalising number with a strong presence in late medieval alchemy and philosophy. Four is the number of rational organisation (Cirlot 1993: 232; Cooper 1990: 115). Four were the principal elements (air, earth, fire, water) in which the prime matter manifested itself; 4, too, were the humours that had to be in perfect balance to ensure human health; 4 are the cardinal points, the arms of the cross, the phases of the moon, the seasons ... and, according to some authors, the main stages in the quest for metal transmutation.

The anchor is not always present, nor is it easy to interpret. This symbol has been strongly entrenched in the publishing trade since it was first used in an emblem by late 15th-century printers of the Aldine Press (Roberts 1893: 218–23), and it persists in the present-day logos and names of some publishers, such as Anchor Books. Its origin appears to have religious or moral connotations, too, as it was often printed together with the motto *Anchora spei* ("anchor of hope"). The Biblical *Epistle to the Hebrews* says: "Which hope we have as the anchor of our soul" (Cirlot 1993: 9). More generally, the anchor appears in religious iconography as a representation of the true faith (Cooper 1990: 12).

Quite significantly, different combinations of single and double crosses, together with the number 4, are extremely frequent in early modern alchemical symbology, to denote such important materials as the crucible, gold, mercury, sulphur, or salt, together with reactions such as distillation (e.g., Anon 1755; Figure 9.2.b). All of these marks are relatively simple, but as the discussion below emphasises, it is difficult to explain their recurrence in religious, alchemical, and other artisanal contexts without offering an explanation that deliberately bridges this series of religious, social, and economic motivations.

The Temple of Solomon

The crucible stamp was, to reiterate the point, a monogram that combined a person's initials with other symbols expressing religious allegiances but also, quite possibly, philosophical ideas that were entrenched in alchemical practice. Although fashion and emulation probably played a role in the spatial and temporal endurance of the mark, nevertheless its use across different media and across innumerable guilds, artists, and workshops in Europe for several centuries demands a more convincing explanation. In fact, a final symbol present in the stamp—the niche-shaped outline often encasing the mark—may partly account for its widespread acceptance.

As noted above, the stamp identified on the exterior crucible bases is often framed by an outline that is square at the base and rounded at the top. Such a shape does not seem incidental, for it is recurrent in masonic books, architecture, and paraphernalia as a representation of the two columns and the vault of Solomon's Temple (Lima and da Silva 2003).

The utopian notion of Solomon's Temple revolves around the Old Testament King Solomon, widely renowned for his wisdom and fairness. Inspired by this figure, Francis Bacon had developed the idea of the temple as a utopian house of wisdom, where knowledge would be generated, applied, and disseminated for the sake of progress, and where human values would be inspired by actual research and tested information. In his own words, published posthumously, Bacon says: "The End of Our Foundation Is the Knowledge of Causes and Secrett Motions of Things; and the Enlarging of the Bounds of the Human Empire, to the Effecting of All Things Possible" (Bacon 1658: 26). Explicitly or implicitly, Bacon's elaboration on Solomon's Temple triggered the foundation of such long-lived institutions as the British Royal Societies or the Ashmolean Museum (Bennett et al. 2000), but the idea was embraced with particular fervour by myriad fraternal institutions that emerged across Europe from the late 17th century onwards—the masonic lodges (Lima and da Silva 2003; Macoy 1989; Figure 9.3).

Notwithstanding an extremely complex internal diversity, innumerable present-day masons (and their detractors) define this institution as "a system of morality veiled in allegory and illustrated by symbols." The historical roots of freemasonry appear to rest in earlier arts and crafts guilds, whose organisational structure and foundational values are mirrored in masonic lodges. Inspired by the ideas of the Enlightenment, the first freemasons were males of moderate to very wealthy social class who gathered together under the avowed premise of mutual respect and protection, defense of honourable values, and anticipation of a better society.

Although some of their foundational myths trace an imagined ancestry back to King Solomon, Hermes Trismegistus, or the Knights Templar, there is no historical evidence for the existence of masonic lodges before the late 17th century (Jacob 2006). It seems plausible that their institutional development relied on medieval structures: As their economic or professional role dissipated, guilds increasingly accepted wealthy laypeople who employed these as structures for their own socialisation (Hamill 1986; Jacob 1991). They often kept preexisting hierarchies of masters, officers, and apprentices and symbols such as the famous compass and square of stonemasons, whilst strengthening the moral commitment of their members.

Many lodges attached a fundamental importance to secrecy, which, in turn, required the use of codes and symbols identifiable by the initiated only. The use of monograms that incorporated several overlapping symbols is common, and these are frequently drawn from religion and alchemy, as both realms incorporate concepts and allegories of idealism, transformation, and perfection. In fact, when one looks at some of the symbols employed by early modern and more recent freemasons, all of the marks identified in the crucible stamp are encountered again (Dennis 2005; Lima and da Silva 2003; Macoy 1989; Roob 2005; Figure 9.3).

Of particular interest is the relationship between the iconography and symbolism of alchemists and freemasons. Although this position was far from widespread (see Principe and Newman 2001), a good number of early modern alchemists saw strong connections between their practical experiments and esoteric and hermetic ideas: The transmutation of matter was used as a metaphor for the transformation of the adept himself, and knowledge was sought in the writings attributed (often erroneously or

Figure 9.3. Detail of the initiation of a freemason, as engraved by Léonard Gabanon in 1745, showing the two columns of Solomon's Temple and other masonic symbols.

deceivingly) to ancient sages. The continued failure of alchemy as a practical discipline to uncover the Philosophers' Stone (the substance or mechanism by which base metal might be transmuted into gold), the increasing marginalisation of more esoteric alchemists from mainstream science, and an accompanying decline in social standing led some to turn to more strongly hermetic endeavours and even to form secret societies (Priesner 2007). The allegory of the alchemist seeking human and material perfection, the secrecy and rarified nature of hermetic work, and the widespread use of symbols understandable only by adepts seemed altogether fitting to some forms of freemasonry (Priesner 2007; Yates 1964, 1975).

When all of these factors are taken into account, it is possible to see the early modern masonic lodge as a synthesis of late medieval institutions such as guilds and philosophical fraternities, incorporating hierarchical structures, initiation rites and tests, secret codes and values, and mutual loyalty. Rather than being an entirely new creation, freemasonry appears as a product of various historical contingencies. Considering these diverse historical backgrounds, one can more easily understand the extreme heterogeneity of the many secret, civic, or ritualised societies that sprang up across early modern Europe, some of which even had little or nothing to do with freemasonry (Clark 2000; Jacob 1991, 2006).

Although the above ideas require much further research and clarification, they provide sufficient background to return now to the main subject of the present chapter: the stamp found in prints and on pots. In its most developed form, this physical product mark seems to have been a symbolic expression of some of the most precious masonic ideals, combining hermetic mysticism and alchemical allegory, and to have spread successfully through the medium of material culture.

Freemasons' Branding?

Previous studies (Hamilton and Lai 1989; Wengrow 2008) and other contributors to this volume have demonstrated that brands have arisen in a wide range of ideological and social settings. For the purposes of this chapter, I will not therefore discuss different definitions of "brand" or elaborate further on non- or pre-Western branding as a phenomenon but would like to recall three characteristics that may reasonably be accepted as defining features of a brand. A successful brand: (1) Draws on human emotions in some structured manner; (2) Often employs communication codes that are different and/or complementary to language; and (3) Promotes consumption patterns that allow mutual recognition among consumers. These points will be briefly developed before I discuss their applicability to this case study.

A branded commodity has to be recognisable. Whether this is achieved through specific sealing, labelling practices, or other means, the consumer needs to be made aware that he or she is acquiring a product with a certain branded identity (Richardson 2008). Such awareness will go beyond the identification of a certain producer or origin (sometimes these may even be concealed or irrelevant); in purchasing such a commodity, the consumer will *feel* that a necessity or desire has been fulfilled, often on a dimension that goes beyond the sheer solution of a practical problem such as the quenching of their thirst or protection from cold temperatures.

Furthermore, brands capitalise on a need for belonging that engages their consumers: Explicitly or implicitly, consumers are perceived to share values, preferences, or lifestyles that are much more meaningful than the mere consumption of the standardised, replaceable commodities that are typically the subject of branding efforts. In fact, the act of consumption is often a celebration of those values and strengthens the ties among those loyal to a brand.

The ways in which these connections among commodities, consumption practices, and patterns of emotion and mutual recognition develop are varied. They rarely emerge from unilinear, self-consciously logical arguments but rather through a network of connections between images and ideas that is often imbued with a great deal of subjectivity. This is probably one reason why ordinary spoken or written language is not the most appropriate strategy for commodity branding: Instead, modern marketing teams, for example, draw on direct sensorial or intellectual associations between commodities and certain objects, peoples, ideas, or experiences.

In many ways, the above ideas would seem to fit quite well the scenario of the marked crucibles as well as that of the books, wares, and other artefacts marked with the stamp at the core of this chapter. The mark would convey strong emotional associations with high moral and civic values; those with sufficient connoisseurship to unpick the subcomponents of the monogram would feel further drawn to acquire those objects instead of the unmarked equivalent, or that of a different producer. Perhaps others would decide against those objects precisely because of those connotations, and not on grounds of quality or price. It seems sensible to assume that knowledgeable consumers of "masonic commodities" would recognise each other mutually as members of a specific social or moral category, whereas others may only have been more casually reassured about the quality and authenticity of such marked products. In other words, the stamped crucible, silver ring, or ceramic pot would promise not only to accomplish their immediate practical roles but also to fulfill broader social or psychological aspirations—and the latter may well have been much more important in determining specific consumption patterns.

The possibility of a brand identity associated with freemason's imagery is consistent with a growing trend, popular in Europe since the late Middle Ages, to consume imagery in a variety of material media. The development of woodblock printing and metal plate engraving in the European Renaissance facilitated the swift spread of printed imagery across different social sectors, leading to a "democratisation in the consumption of imagery" (Gaimster 1997: 142 [see also pp. 37–40, 126–36, and 142–55], 2003). During the 15th century, craftsmen incorporated some of this printed imagery into different media, including sculpture, medals, metal vessels, wooden furniture, domestic furnishings, and pottery. In the upheaval of the Reformation, religious and political images inspired by print media became recurrent in domestic and other objects. Stoneware potters, for example, applied moulded motifs to their productions, representing biblical episodes, satirical caricatures, moralising scenes, and political portraits and slogans (Gaimster 1997: 142–55).

Thus, ceramics became a more or less explicit channel for the communication of values, beliefs and bonds. The images utilised were probably the result of each potter's individual preferences, but they must have found a market of people from all social levels who wished to display their own leanings in the religious and political arena. As a matter of fact, the overt Protestant allegiance of some trade guilds led them to migrate to regions where they found more sympathy. Not only stoneware vessels, but also devotional pipeclay figurines and glazed earthenware stove-tiles conveyed blatant political and religious messages in the form of images taken from prints as well as from ecclesiastical altarpieces and statuary (Gaimster 2003). All of these were ubiquitous artefacts that made it possible for anyone to celebrate their personal beliefs in a visible manner, even those who were hardly literate.

Against this background, the possibility of freemasons—an influential, confident, and affluent social segment—consuming cherished imagery as part of physical goods made in a range of media does not seem surprising. Any loose labelling of this practice as proto-Western branding, however, is not without its problems. First, the deliberately hermetic semantics of the symbol would seem at odds with what we might construe as conventional modern branding strategies. The fact that an 18th-century writer (see above) is bewildered by this stamp at the same time as these marked commodities are being produced demonstrates this point. Such secrecy and hermetism are at the core of freemasonry, thus they may have made these commodities particularly appealing to the masons' taste. Nevertheless, one is left wondering the extent to which non-masons would take notice of the mark.

Similarly, one may assume that a master crucible maker would originally be acquainted with the meaning of their own house mark (but see below for whether this is true for the makers of becherovka, a traditional herbal drink). However, would the potters working for them be aware of this? Allegedly, this would only be the case if the potters were also initiates who thus shared the beliefs associated with the secret symbol. In this sense, we should be cautious before assuming connections between masonic commodities in the archaeological and historical record and the spread of actual masons, masonic lodges, or masonic ideas.

Of relevance to this point is Bevan's discussion of the different ways in which equally branded commodities may change hands and meanings in the market; in particular, the "leakage" of branded commodities and their ideological associations, which highlights the active role that consumers may have in the constant redefinition and distribution of branded goods (Bevan, this volume). It seems obvious that the goods marked with the stamp discussed in this chapter could have evoked very diverse aspirations among different social sectors, ranging from the overt commitment to specific ideologies, to the mere recognition of a high-quality product and its reassuring labels, to positions of complete indifference.

Another contentious issue is the occurrence of this potential form of marking or branding in such a wide range of media. As mentioned above, variations of this stamp were used by crucible makers, but also by potters, silversmiths, merchants, engravers, and many other craftspeople across Europe, who had no institutional or commercial connections among them. This was not a single company branding a range of products, just a wide array of producers who may not have even known of each other, but who nonetheless employed similar marks. Such a realisation again emphasises the possibility that a very similar stamp might have conveyed different ideas in different contexts. Moreover, in some instances at least, the stamp could be an expression of the producer's values or allegiances much more so than an appeal to the consumer's emotions. If this was the case, perhaps we might speak of patterns of sponsorship in addition to or instead of branding. In fact, the influence of the masonic lodges may have been more marked at the manufacturing rather than at the consumption end of the economy, especially in light of the historical entanglement of professional guilds and masonic lodges discussed above.

Finally, a related issue that should not be overlooked is the extremely long temporal span during which this mark was used. Similarities between 15th-century and 19th-century monograms combining crosses, anchors, and the number 4 cannot be ignored—but it would be erroneous to use these similarities to draw any easy conclusions. It is particularly important to note that the use of this mark starts well before any documented example of a masonic lodge, and thus labelling all of these as "masonic stamps" would be misleading. A relevant case is that of the Kaufmann family of Obernzell, likely the surname behind the letter "K" frequently found in these crucible stamps (Figure 9.2): In the 17th century, Gabriel Kaufmann enjoyed the commission of supplying crucibles to the Imperial Mint (Bauer 1983: 31). Two centuries later, we find a likely descendant, Josef Kaufmann, associated with Gg. Saxinger and producing crucibles with remarkably similar stamps. Were the later generations preserving the illustrious stamp as a sign of tradition or reputation, and were they aware of its particular significance? Despite the potential links between the organisation and culture of medieval guilds, alchemical laboratories, and modern masonic lodges, such connections need to be further questioned and tested rather than generalised.

The above concerns are usefully revisited via one final example: the modern bottle of becherovka. Becherovka is an herbal alcoholic drink that has been produced in the Czech Republic for the last 200 years. It was originally sold as a medicinal tonic by its inventor, the pharmacist Jan Becher, but today the becherovka brand belongs to the multinational Pernord Richard, and the spirit is sold to tourists and locals alike as the Czech "national drink." The consumption patterns and social connotations of this liquor have changed substantially, but the marketing team has made an effort to preserve its traditional appearance—for example, by keeping Jan Becher as a trademark and, just as importantly, by maintaining the house monogram: one that, incidentally, combines a number 4, a double cross, and what could be seen as a stylised anchor (Figure 9.4). I do not know whether present-day makers of becherovka know the history of their housemark or how many drinkers notice or care about the stamp. In any case, any sweeping connection to alchemy and secret societies seems highly unlikely, but the overall effect of the mark is still arguably to invest the product with a strong sense of craft lineage and, however indirectly and/or subconsciously, to associate it with other seemingly authentic products, marks, and concepts wrapped up in centuries of European economic, religious, and social life.

Conclusion

In a recent paper, Richardson (2008) has made a compelling argument highlighting the role of medieval guilds as "branding mechanisms" that encouraged or enforced the manufacture of standardised, high-quality, recognisable products as means of protecting established products and industries from outside imitation and indirectly encouraging the economic development of certain towns and/or regions. The present study has shown that relatively inconspicuous images such as potter stamps, house marks, and other monograms may also have provided a subtle medium for the expression and celebration of ideas and allegiances and that, in certain cases, these images may have been forms of commodity branding that drew on particular emotional responses to certain subjects and promoted specific consumption patterns. More precisely, it was suggested that the marks stamped on Bavarian crucibles and found on an array of other early modern material culture synthesised a set of elevated ideas about contemporary religious belief, alchemical allegories, and, at least in its later forms, masonic codes.

Although the above analysis emphasises the interconnectedness of these issues in medieval economic and social life, I would end by stressing that this should not encourage the writing of conspiracy theories (masonic or otherwise) into the historical record. As a matter of fact, it seems impossible to trace a single agent explaining why and how people have used these marks or responded to them, which underscores the lack of a single "corporate" effort as generally identifiable behind modern branding ambitions.

At one level, I hope I have shown that material culture can reveal information about the transmission and development of ideas: In particular,

Figure 9.4. A modern bottle of becherovka. Note Jan Becher's monogram both moulded on the top part of the glass and reproduced in the centre of the label. The modern reproduction on the label has lost the second horizontal crossbar, still present in the moulded version.

traditionally obscure topics such as the early modern translocation of alchemy into secret and esoteric societies, the sociological and ideological foundations of masonic lodges, and the links between these and professional guilds, may be illuminated from this renewed perspective. At another level, the concept of commodity branding emerges as a useful touchstone with which to address patterns of production and consumption in the archaeological record.

However, we should be wary of the risks of oversimplification in such approaches. Tempting as it may be to conflate "branded" goods in the archaeological record with more modern structures of commodity branding, further diachronic perspectives are likely to show that the philosophy, market, and consumption patterns behind seemingly identical, long-lived, and physically marked goods are, in fact, more diverse than initially apparent. As the super-imposed, often highly obscure components of the crucible marks attest, the degree of literacy with which a consumer reads a brand image varies enormously, and (variously to the despair or to the advantage of modern marketing gurus), consumers may purchase (or reject) brands with very selective awareness of the sophisticated emotions, lifestyles, and mutual aspirations allegedly ascribed to them.

Note

1. This work has only been possible through the goodwill and generosity of a number of individuals who sustained my curiosity by referring me to obscure sources and museum collections, or by considering and discussing my shapeless ideas as they developed. I am particularly indebted to Ingolf Bauer, Lesley Bushnell, Karl Friedl, David Gaimster, Paulo Dordio Gomes, Rudolf Hammel, Thilo Rehren, and Sigrid von Osten as well as to the participants of the Cultures of Commodity Branding Conference. I am also grateful to Andy Bevan and David Wengrow for inviting me to join their original venture—especially to Andy for his excellent editorial advice. I would also like to thank Louise Iles for kindly redrawing the stamps in Figure 9.2. This research was mostly carried out on the side while I was enjoying funding from the AHRC, Caixanova Foundation, and the University of London Central Research Fund.

References

Anon. 1755. *Medicinisch-Chymisch- und Alchemistisches Oraculum*. http://www.levity.com/alchemy/alchemical_symbols00.html (accessed 7 March, 2008).

Bacon, F. 1658. *New Atlantis, A Work Unfinished*. London.

Bauer, I. 1983. *Handbuch und Führer zum Keramikmuseum Schloß Obernzell, Zweigmuseum des Bayerischen Nationalmuseums*. Munich: Bayerisches Nationalmuseum.

Bennett, J. A., S. A. Johnston, and A. V. Simcock. 2000. *Solomon's House in Oxford: New Finds from the First Museum*. Oxford: Museum of the History of Science.

Black, A. 1984. *Guilds and Civil Society in European Political Thought from the Twelfth Century to the Present*. Ithaca, NY: Cornell University Press.

Cirlot, J. E. 1993. *A Dictionary of Symbols*. London: Routledge.

Clark, P. 2000. *British Clubs and Societies 1580–1800: The Origins of an Associational World*. Oxford: Clarendon Press.

Cooper, J. C. 1990. *An Illustrated Encyclopaedia of Traditional Symbols*. London: Thames and Hudson.

Cotter, J. P. 1992. "The mystery of the Hessian wares": Post-medieval triangular crucibles. In D. Gaimster and M. Redknap (Eds.), *Everyday and Exotic Pottery from Europe c. 650–1900. Studies in Honour of John G. Hurst*: 256–72. Oxford: Oxbow.

Dennis, V. S. 2005. *Discovering Friendly and Fraternal Societies: Their Badges and Regalia*. Malta: Shire Publications.

Epstein, S. 1991. *Wage Labor and Guilds in Medieval Europe*. Chapel Hill: University of North Carolina Press.

Fisher, J. 1521. *Contio Quam Anglice Habuit Reverendus Pater Ioannes Roffensis Episcopus in Celeberrimo Nobilium Conuentu Londini...* . Cambridge: Ioannem Siberch.

Gadd, I. A. and P. Wallis, Eds. 2002. *Guilds, Society and Economy in London, 1450–1800*. London: Centre for Metropolitan History, Institute of Historical Research, Guildhall Library.

Gadd, I. A. and P. Wallis, Eds. 2006. *Guilds and Association in Europe, 900–1900*. London: Centre for Metropolitan History, Institute of Historical Research—School of Advanced Study, University of London.

Gaimster, D. 1997. *German Stoneware 1200–1900: Archaeology and Cultural History*. London: British Museum Press.

Gaimster, D. 2003. Pots, prints and Protestantism: Changing mentalities in the urban domestic sphere, *c*. 1480–1580. In D. Gaimster and R. Gilchrist (Eds.), *The Archaeology of the Reformation 1480–1580*: 122–44. Leeds, UK: Maney.

Girling, F. A. 1962. *English Merchants' Marks: A Field Survey of Marks Made by Merchants and Tradesmen in England between 1400 and 1700*. London: Lion and Unicorn Press.

Hamill, J. 1986. *The Craft: A History of English Freemasonry*. Leighton Buzzard, UK: Crucible.

Hamilton, G. and C. Lai. 1989. Consumerism without capitalism: Consumption and brand names in late imperial China. In H. Rutz and B. Orlove (Eds.), *The Social Economy of Consumption*: 253–79. Lanham, MD: University Press of America.

Hoover, H. C. and L. H. Hoover. 1950. *Georgius Agricola: De Re Metallica*. New York: Dover Publications.

Jacob, M. C. 1991. *Living the Enlightenment. Freemasonry and Politics in Eighteenth-Century Europe*. New York: Oxford University Press.

Jacob, M. C. 2006. *The Origins of Freemasonry: Facts and Fictions*. Philadelphia: University of Pennsylvania Press.

Jones-Baker, D. 1996. Makers' marks engraved in graffito on English medieval and post-medieval tombs and other funeral monuments. *The Antiquaries Journal* 76:254–57.

Lima, T. A. and M. N. da Silva. 2003. Alquimia, ocultismo, maçonaria: O ouro e o simbolismo hermético dos cadinhos (séculos XVIII e XIX). *Anais do Museu Paulista* 8/9:9–54.

Macoy, R. 1989. *A Dictionary of Freemasonry: A Compendium of Masonic History, Symbolism, Rituals, Literature, and Myth*. New York: Bell.

Martinón-Torres, M. 2005. Chymistry and crucibles in the Renaissance laboratory: An archaeometric and historical study. Ph.D. thesis, University of London.

Martinón-Torres, M. 2007. The tools of the chymist: Archaeological and scientific analyses of early modern laboratories. In L. M. Principe (Ed.), *Chymists and Chymistry: Studies in the History of Alchemy and Early Chemistry*: 149–63. Sagamore Beach, MA: Science History Publications and Chemical Heritage Foundation.

Martinón-Torres, M., I. Freestone, A. Hunt, and Th. Rehren. 2008. Mass-produced mullite crucibles in medieval Europe: Manufacture and material properties. *Journal of the American Ceramic Society* 91:2071–74.

Martinón-Torres, M. and Th. Rehren. 2005. Alchemy, chemistry and metallurgy in Renaissance Europe. A wider context for fire assay remains. *Historical Metallurgy* 39:14–31.

Martinón-Torres, M. and Th. Rehren. 2009. The production and consumption of crucibles in the post-medieval world: A study of materials and materialities. *Archaeometry* 51:49–74.

Martinón-Torres, M., Th. Rehren, and I. C. Freestone. 2006. Mullite and the mystery of the Hessian wares. *Nature* 444:437–38.

Milne, G. 2003. *The Port of Medieval London*. Stroud, UK: Tempus.

Priesner, C. 2007. Alchemy and enlightenment in Germany. Ideas, biographies, secret societies and a changing cultural context. In L. M. Principe (Ed.), *Chymists and Chymistry: Studies in the History of Alchemy and Early Chemistry*: 253–64. Sagamore Beach, MA: Science History Publications and Chemical Heritage Foundation.

Principe, L. M. and W. R. Newman. 2001. Some problems with the historiography of alchemy. In W. R. Newman and A. Grafton (Eds.), *Secrets of Nature: Astrology and Alchemy in Early Modern Europe*: 385–431. Cambridge, MA: MIT Press.

Richardson, G. 2008. *Brand Names before the Industrial Revolution*. NBER Working Paper No. 13930. Cambridge, MA: National Bureau of Economic Research.

Roberts, W. 1893. *Printers' Marks: A Chapter in the History of Typography*. London: George Bell and Sons.

Roob, A. 2005. *The Hermetic Museum: Alchemy and Mysticism*. London: Taschen

Stephan, H.-G. 1995. *Großalmerode: Ein europäisches Zentrum der Herstellung von technischer Keramik. Die Geschichte der keramischen Gewerbe in Großalmerode und Epterode und die Entwicklung ihrer Produktion vom 12. bis zum 19. Jahrhundert. Teil II.* Großalmerode, Germany: Glas- und Keramikmuseum Großalmerode.

Wengrow, D. 2008. Prehistories of commodity branding. *Current Anthropology* 49:7–34.

Yates, F. A. 1964. *Giordano Bruno and the Hermetic Tradition*. London: Routledge and K. Paul.

Yates, F. A. 1975. *The Rosicrucian Enlightenment*. St. Albans, UK: Paladin.

Chapter 10

The Second-Hand Brand

Liquid Assets and Borrowed Goods

Alison J. Clarke

Over the last decade, a body of interdisciplinary scholarship has arisen in response to a perceived overemphasis on the empirical and theoretical significance of first-phase consumption. Waste, decay, disposal, dispersal, ridding, and second-hand markets have taken on a new significance within both social sciences and historical approaches to consumption and material cultures (Clarke 2000; DeSilvey 2006; Gregson 2007; Crewe and Gregson 1998; Gregson and Crewe 2003; Hetherington 2004; Lucas 2002; Reno 2009). Despite the timely expansion of research beyond purchase, display, and conventional models of ownership and consumption, modern brands (as marked forms of material culture belonging to an exceptional hierarchy of value within first-phase consumption) have received minimal scrutiny with regard to the particularity of their dispersal.

Prior to the present-day widespread access to open digitalised sites of consumption (such as on-line auctions), used and "out-of-season" goods, from clothing to furniture, circulated in markets managed almost exclusively via expert mediators or direct face-to-face encounter (Edwards 2005; Lemire 2004). The growth of global trading sites (such as eBay) has wrenched second-hand goods from the networks of expert traders into the risk-filled and unbounded realm of public, informal connoisseurship on a mass scale.[1] The contemporary dispersal of used goods is no longer tied to any "logic" of expertise or mechanisms of remerchandising. Contrary to Baudrillard's notion of the brand as an ephemeral sign offering fleeting, "short-term gratification" (see Wengrow this volume; Baudrillard 1996 [1968]) in the discordant, on-line market of second-hand-wares, the brand takes on a new "virtualised" significance of endurance and solidity; it can be easily recognized, identified, and tracked.

Although the value of archetypal globalised brands has long been understood as generated through "the prism of the local" (Clarke 2000; Miller 1995), in the informal economy of the on-line auction, the ambivalences of a brand's value are simultaneously globally negotiated and locally contested. From assessing the risks of counterfeiting to testing the social currency of specific designs, the on-line encounter is composed of transactions that mediate the enormity of a potentially worldwide audience with the micro-levels of local product cultures.

This chapter explores second-hand brand economies, as viewed through the phenomenon of on-line consumption, to challenge the assumption that used brands simply mirror the logic of first-phase brand consumption or automatically dissipate in value. Instead, it asserts that through the interactive social and collective processes of on-line auction specific used brands actually accrue more value than their first-hand counterparts, whereas others are made to depreciate significantly. Beginning with an exploration of a growing move towards "lease ownership" (in which brands are literally "hired out"), the chapter reveals how the commercial market itself is beginning to incorporate the spectre of second-hand brand value into its rhetoric and aesthetic. Through ethnographic detail of on-line auction purchase, it goes on to explore the practices of riddance and dispersal that underpin the making of value around competing second-hand brands.

The chapter concludes that the secondary market for branded goods has become so influential that brand strategists and economists are being forced to incorporate the second-hand brand economy into their overall brand development; first-hand brand identities are becoming swamped in the significance of their ghostly counterparts lurking in the unpoliced virtual worlds of post-retail consumption. The actual practices of design and styling, product development, and research are shifting as the second-hand market asserts a backward influence over the creation of the branded good itself. The practice of design, once understood as a practice of innovation, is now recast as one of consolidation in which the "whole life" of the product (from production through to dispersal and decay) is acknowledged. Designers are increasingly forced to incorporate and respond to the post-primary consumption cultures of products, with the discernible consequence that anthropologists and ethnographers have become commonly incorporated into the branding process itself as "brand archaeologists" (Clarke 2010; Drazin and Garvey 2009).

Liquid Assets and Borrowed Brands

Unanchored from their first-phase mediations (retail space, advertising, product placement, celebrity endorsement, brand management) and uncoupled from their original socioeconomic and geocultural contexts, constellations and "diasporas" of second-hand brands occupy an arguably "transcendent" realm as commodities. But how are such branded goods disposed of in relation to those of their unbranded counterparts? How do the mechanisms of first-phase brand mediation impact on the second-hand economy? And what implications does the new visibility of this "brand-underworld" have for our understanding of brand consumption?

Although the temporalities of brand value are made quantifiable through the trajectories of on-line auction, brand strategists are all too aware of their inability to control the "brand leakage" that generates the alternative economies of value into which their brands fall (as "monetizable symbolic values"; Gorz 2003: 60).

Assumptions regarding the value of newness, innovation, and novelty, the linchpins of modern capitalism and product development, are thrown into relief by the dynamics of a previously obscured aspect of modern consumption: the thriving second-hand market. Models of brand ownership, it could be argued, are moving away from models of possession, increasingly resembling instead those of leasing, borrowing, and recycling.[2] These developments in brand acquisition and circulation echo the theories of political ecologists of the late 1970s (Gorz 1980; Papanek 1985 [1971]) who consistently argued for the necessity of radically alternative dematerialised relations to product consumption as the path to a sustainable economy. The following example of on-line brand consumption explores how the phenomenon of the "after-life brand" interconnects with the mechanisms of a more fluid approach to commodity ownership and new market practices that promote brands as "liquid assets" rather than owned goods; a process that strips away the commodity's "immaterial assets" (Arvidsson 2005), leaving the brand "enacted" but unconsolidated as an absolute purchase.

Under the slogan "Guilty pleasure—minus the guilt," the popular on-line fashion site Bag Borrow or Steal™ offers its members the possibility of renting authentic designer branded luxury handbags and accessories. Delivered to the front door in gift-like packaging, the "as-new" condition items can be rented weekly (a small item costing $100), monthly (at $300), or indefinitely (the so-called steal option), whereby the full cost of the item is debited to the member's credit card. To facilitate the closest possible adherence to the nuanced changes of fashion, the leased items can be replaced by return mail and traded up for a newer "must-have" model within 24 hours. After successfully fulfilling a six-month membership test period, clients are permitted an upgrade allowing access to genuine couture items. A VIP membership option enables customers to join an exclusive waiting list for the most sought-after designs from labels such as Mulberry, Prada, Louis Vuitton, Fendi, and Chanel, the brands familiarly seen hanging from the arms of celebrities in the pages of fashion/gossip magazines or so-called fabloids.[3] Customers are seamlessly incorporated into the fabloid world of minutely observed bodily and fashion change and lent celebrity-like status by usurping the retail line of highly coveted "it" items.

Bag Borrow or Steal™ celebrated its relation to the rarefied world of transient designer fashion brands through its product placement in the *Sex in the City* movie (2008), a spin-off from the popular U.S. television series renowned for its orgiastic consumption of must-have brand items, from designer shoes to baby buggies, that are interwoven into the manners and mores of women's on-screen urban lives. The historically intertwined relation of glamour and branded luxury goods, from Hollywood through to fabloids, is a well-documented phenomenon. As a collective discourse, glamour and its brands, it has been asserted, specifically touches those "on the margins of society on account of their economic position, ethnicity or sexual orientation" (Gundle 2008: 356).

The potentially transgressive power of glamour and associated images and brands is an established tenet of feminist discourse (Modleski 1986; Peiss 1998). Whilst relying on an established fashion system in which media penetration of social relations is paramount, the leasing method of Bag Borrow or Steal™ appears to offer a potentially democratised relation to fashion consumption in which economic power is not directly aligned with brand access. The company encourages the indulgences of the fashion-driven compulsive on-line credit card shopping (Park and Davis 2005) and the thrift of deferred gratification; the hiatus between longing and owning that came to define modern consumerism (Campbell 1987). That customers can "borrow" brand salience and then casually retrade it suggests a newly evolved level of sophistication and knowledge that underpins an ever faster-paced fashion industry (O'Cass 2004) and a burgeoning on-line fashion trading culture (Brown et al. 2008; Park et al. 2007).

As a conduit for fashion advice that simultaneously celebrates and neutralises brand power, Bag Steal or Borrow belongs to an established genre of fashion advice. Since the 1970s, a highly successful product-based style consultancy Colour-Me-Beautiful (CMB) has offered, under the slogan "Colour Me Confident," a paid service aimed predominantly at a middle-aged audience otherwise precluded from "high fashion" knowledge. By supplanting the accelerated seasonal style-change of the fashion industry and its brands with a colour-led approach to individual style, CMB "tames" the vagaries of fashion and subjugates associated social anxiety over fashion choice (see Clarke and Miller 2002; Grove-White 2001).

Advisory retail methods such as CMB rationalise the often bewildering, fast-paced fashion industry and filter fashion choice through collectively negotiated expertise (comparable to the consensus generated through on-line auction). At the high-end of the market, the Bag Steal or Borrow website similarly acts as secure filter for optimising the "right" fashion choices, but it also authenticates items, in terms of provenance and cultural capital, and most significantly—appears to beat the highly developed and notoriously ruthless branded fashion industry at its own game. "It" bags and accessories are extracted from the obfuscated mayhem of fashion mediation and consumption—and delivered quite literally gift-wrapped to the door.

The membership scheme offers added value by transforming the barefaced branded commodities of a fickle fashion industry into intimate inalienable gifts (they are shared and circulated in a cognitive and haptic sense within the inner circle of like-minded fashion-savvy members). The "hot" brands arrive "new-like" but with the added mystique of a glamorous provenance; which party or society event the selected handbag previously attended—the recipient can only ponder. Unlike the fortuitous "found" object from the second-hand bazaar, the "borrowed" handbag bears the mysterious patina of use by a selected style elite (see Bevan, this volume, for comparative historical example of brands and elites). The customer has the possibility of managing the brand and basking in the aura of its associations, whilst avoiding the negative aspects of a short-lived "splurge" purchase (Fitzmaurice 2008). This form of consumption signals a shared acknowledgement of the futility of possession in a culture of brand satiation and a collective effort to tame the endless turnover of styles.

This example of brand mediation might at first glance seem aberrant of the otherwise mundane culture of on-line shopping, but its pretext offers insight into a potential shift in the consumption of designed goods and our understanding of branded commodities themselves.

Issues of "brand equity," rather than the materiality of the objects themselves, are the mainstay of business literature (Aaker 1991, 1996; Schmitt 1999) as concern with the brand has moved from consumer marketing to management. However, within the same literature the origins and underpinnings of brand equity remain vague; despite the now quantifiable presence of brand equity in shareholders' calculations of predictable future earnings, its presence is mostly attributed to "diffuse factors like tradition, coincidence or luck" (Arvidsson 2005: 239). An established tenet is, however, the role of brand managers in generating, controlling, channelling, and positioning these "immaterial assets" in increasingly diverse and complex circumstances of consumption. For "the brand," observes Arvidsson in his critical overview of brand cultures, "refers not primarily to the product, but to *the context of consumption*. It stands for a specific way of using the object, a propertied form of life to be realized in consumption" (2005: 244). The intervention of brand strategists in shaping the "immaterial labour" (Arvidsson 2005) of consumption that surrounds brands is well documented (Frank 1997; Stalnaker 2002): "Stealth," "viral," and "guerilla" marketing; "word-of-mouth" grass-roots advertising; and trend-spotting anthropologists "injecting" brands into "cool" urban spaces are all familiar themes of urban myth and academic research alike. Although there are folkloric tales of the Mercedes corporation systematically buying up their used cars to prevent brand penetration in lower income and less salubrious neighbourhoods,[4] the general market in used goods outside the car industry has not presented itself to brand managers as a tangible entity worthy of serious concern.

More recently, however, brand strategists have become increasingly concerned with the phenomenon of lease ownership and the growing significance of a buoyant, unregulated second-hand market in used branded goods. In *Futureshop*, a best-selling retail management book, marketing advisor Daniel Nissanoff (2006) warns retailers and brand managers that resale value is beginning to migrate and become incorporated analytically as a component of people's purchase.[5] Brand perception, in other words, is becoming quantifiable. After decades of research and billions of marketing dollars spent on notions such as "emotional engineering," "haptic branding," and "product semiotics," figures like Nissanoff warn that brand makers have become dangerously vulnerable to the anarchic underbelly of primary purchase in the form of a proliferating and an increasingly sophisticated consumer second-hand market. The on-line informal market in branded goods, warn brand strategists, is now arguably as significant in generating brand identity as top-down brand management. Viewed as "liquid assets" rather than possessions, consumers glean optimal abstract value from the branded good and then rerelease it into the market to recoup a quantifiable monetary value and re-invest accordingly. Why *possess* a luxury-branded handbag when you can simply share it and forego the social anxiety of being left with an expensive, redundant, and out-moded model? In this lease fashion system, the concept of price is dislocated from the material commodity and attached to a consensually generated measure of value ascertained through the actual processes of borrowing and dis-owning.

Riddance and Dispersals: The Second-Hand Brand on eBay

The significance of the second-phase market and its relationship to specific genres of material culture and modes of possession is well documented in the broader terms of consumption history. Hinton's (2002) study of 16th-century Venetian auctions, for example, reveals the second-hand economy as key to establishing the value of people and things in the Renaissance period; it is through the moveable wealth of households, the flow and trajectories of specific types of objects from beds to ceramics, that familial hierarchies and kinship is made. In Stallybrass's (1998) often-cited polemic on the commodity form, the transience of everyday material life in Karl Marx's household (as objects ranging from Marx's own overcoat to the engraved family silver pass in and out of the pawn shop) reveals "possession" to be a fleeting and tentative concept in 19th-century British households.

An established body of historical literature addressing the issues of second-hand markets and household divestment is matched by a more recent turn to issues of disposal, divestment, and riddance in the sociology and archaeology of contemporary consumption (see Buchli and Lucas [2001], Rathje and Murphy [2001], and Scanlan [2005] on rubbish and garbage; Hawkins and Muecke [2003] on culture and waste, Norris [2004] on cloth recycling; Gregson [2007] and Hetherington [2004] on second-handness, consumption, and disposal).

Yet in these accounts, the branded good is entirely absent as a discreet form or genre of material culture. It is as if, through the processes of domestication and divestment, the first-hand commodity aesthetic that originally "marked" the branded commodity, has been rendered impotent. Although different materialities and cultures of goods—from furnishing to toys, plastic bottles to jam-jar lids—generate different trajectories of ridding and dispersal, the branded commodity is largely absent from this growing literature on divestment. There appears to be an assumption that branded commodities, as marked entities, without their salience and post-consumption, are neutralised and stripped of their enchantment; they are fragile, ephemeral, fleeting entities that, unhinged from the transcendent moment of first-brand purchase and no longer "brand" new, are democratised among the heterogeneous piles of household flotsam.

In contrast to the face-to-face relations of second-hand markets and car boot (trunk) sales, the anonymity of the on-line auction has evolved as a key component of the previously invisible ridding and sorting processes of the household. The term "eBay," as used by the informants and respondents in this research, refers to both eBay the firm and eBay as a set of everyday cultural practices. Freed-up from the historical ties, localities, and personal relations that are typically depicted in the established ethnographic literature around second-hand markets (for garage sales, see Hermann and Soiffer [1984]; for site-based auctions, see Smith [1989]; for ethnically bound trading fairs, see Stewart [1992]), the exponential growth in on-line second-hand trading has propelled branded commodities into a heterogeneous, virtual "bazaar economy."

Unlike the relative safety of the reciprocal and normative cultures generated in garage sales, nearly new sales, site-based auctions, and horse fairs, the hazardous unpoliced, free-market on-line auction sites offer minimal and informal mechanisms (such as notice boards and community forums) to limit hazards ranging from the purchase of fakes to being duped by bogus sellers and bidders (Soteriou 2007).. Awash with millions of items (48,765 women's accessories were available on the U.K. eBay site alone at the time I was writing this chapter), brand is a principal means of navigating eBay space. So redolent is the brand to the functioning of the auction that the corporation has recently been forced to outlaw the use of bogus brand-attribution (i.e., "For sale–*not* Versace dress"), habitually used as a means of creating a "false trace" amongst the millions of otherwise indistinguishable items.

In this bazaar economy, brands are used to mark out virtual spaces of bounded interaction. Fake or authentic, used or new, brands lie at the core of eBay's commercial success as one of the world's biggest on-line retailers. A succession of high-profile legal cases regarding brand protection (that led to contradictory judgements in the United States and France over questions of liability for the sale of counterfeit goods on the web) led to a recent initiative, dubbed the "counterfeit summit," bringing together eBay lawyers and brand representatives keen to stymie the burgeoning on-line trade in counterfeit brands (see Kleinman 2008; Rosencrance 2008). The eBay firm, aided by the legal ambiguity surrounding the question of where responsibility falls for policing counterfeit goods, has consistently resisted calls by luxury and designer-brand firms such as LVMH and Walpole to enforce more rigorous controls on its virtual marketplace.[6]

eBay founders strongly defend the definition of their enterprise as a neoliberal, free-market trading space underpinned by "fundamental values" (Cohen 2002: 28) and individual community participation that appeals to what Robins and Webster define as a type of "techno-communitarianism" that attempts to "reconcile political idealism with the corporate reality principle" (1999: 223).

However, so endemic is the counterfeiting on eBay that the on-line auctioneer is viewed by many as a major impetus behind the counterfeiting of most top-end luxury goods, from Jimmy Choo shoes to Fendi clutch-bags. The same "fabloid" economy of desire for up-to-the-minute style items glimpsed in the lease economy of Bag Steal and Borrow on-line rental service is seemingly generating an anarchic underworld of brand trading in which the authentic and the fake are loosely interchangeable as long as they serve the function in marking the fleeting nuances of style change. But the protectionism of the luxury branding groups is aimed, not just at maintaining intellectual copyright, but of preventing what has been described as the "chaving" of luxury brands, and, more specifically, "the appropriation of signature products by fast-money and down-market celebrities, and even working-class subcultures" (Gundle 2008: 337).

Brands in secondary circulation do not, however, merely glean their value from the formal mechanisms of glamour. Rather, they become embedded in narratives enacted as part of the performative eBay culture and take on various rhetorical forms tied to specific second-hand brands.

These rhetorical narratives, although arguably representative of what Arvidsson (2005: 242) describes as the "community" of brand experience that marketing strategists are keen to invoke, take on an entirely different guise in this vast, indiscriminate "people's archive" in which objects of history, nostalgia, and longing are inseparable (as a lived experience) from those with brand identity. ebay serves as a perpetual, collectively owned archaeological site of second-hand brands.

A Bazaar of Pre-Loved Brands

In his extensive study of on-line trading practices, Hillis has argued that spaces such as eBay offer a collective forum for retrieving and reviving memory, a collective archive of everyday material culture in which "the past" becomes a material object, and "[eBay] an endless documentary record available for present-day personal appropriation" (Hillis 2006: 141). Although it is, in fact, eBay auto sales that generates the largest percentage of eBay sales, the company promotes "collectibles" as "the heart and soul of eBay," thus appealing to what Hillis describes as "the fetishized Western desire to acquire under the rubric of collecting what previously was missing or lost" (p. 142).

The deliberate articulation of the company's relation to memory, argues Hillis, is merely an extension of the media-friendly, human-interest aspects of eBay that serve to increase the overall site activity. These include the infamous grilled cheese sandwich purportedly etched with the face of the Virgin Mary, which sold to an Internet casino for $28,000 USD, and the case of the embittered parents who sold off their children's PlayStations on Christmas eve at an enormous profit (as bidders sort to collectively punish the children for their undisclosed naughtiness). Both stories having achieved the status of urban myths largely through their appeal to the narrative encircling the traded items and the stark way in which everything and anything (ridiculous or sacred objects, children's Christmas gifts) can be instantaneously released into a brutal marketplace.

Exchange and use values negotiated on eBay invoke elaborate, at times sensational, narrative histories contextualising an object's past—and it is this process that actively transforms immaterial commodities into highly coveted items. This may take the form of a simple story describing how one acquired a particular item, a vintage labelled jacket from a Paris boutique or an Ercol 1950s table from an estate sale of a rich farmer. The narrative treads a thin line between enhancing the saleability of the object or rendering it inalienable. Furthermore, these narrative devices adhere to local styles and repertoires that are sensitive to very specific understandings of commodity cultures. Selling the wedding dress of an unfaithful former wife found in the attic after she hastily moved out may be hilariously funny in the United States (modelled by the abandoned husband, the item sold for over $6,000), but might not translate well, for example, to an Austrian eBay where the market is in its early stages and its "economy of fun" is yet to be established. It is the narrative effect of these spatialised performances, Hillis argues, that "exceeds the transactional logic that comprises the commonsense understanding of eBay as a giant garage sale, swap meet or old-fashioned auction house writ virtual" (Hillis 2006: 142).

In this narrative context in which all bidders, and even watchers and browsers, contribute to the story line and the narrative device, the "true" story, image, or provenance of an item is elusive. The "situatedness" of brands coupled with an everyday (rather than spectacular) subjective narrative that confers testimony of social salience makes the ideal eBay commodity. This tentative coming together of brand value and subjective narrative is visible in the process of resettling unwanted items. Listings can read as apologies, even eulogies, to the discarded object as witnessed in the following extracts from eBay sellers;

> My mother bought me this Kenwood food-mixer as a wedding present. It is hardly used (sorry mum!) and we are sad to let it go—I'm sure it makes great bread—but with a new born and a dog to look after we don't have time to use it! (eBay.co.uk/home & garden/14 July 2009)

> You are bidding on my son's Maxi Cosi reclining car seat. He has grown out of it and as much as he loved to kip in it, it is now time to move on to a booster seat ... [signalling contemplative pause]. We have never had a problem with this car seat and it has never been in an accident. It has come from a smoke free and pet free home. (eBay.co.uk/baby/car seats 15 July 2009)

> The fabric [of this Maxi Cosi] is black in colour with some dark grey. There are no stains and tears. There are a few marks on the belt strap, as shown in the photo, unavoidable with small children I feel! (eBay.co.uk/baby/car seats 15 July 2009)

Attaching too much emotional significance to an item (mentioning death, ill fate, bad experiences, etc.) is a hindrance to sale—in this respect, it is *branded* commodities that best circumvent the overly subjective stories of individual sellers often required for nonbranded goods that need greater confirmation of their provenance. It is not collectibles, as promoted by eBay the corporation, that fulfill the eBay founder's idyll of a "friction-free-market" but branded commodities that provide a "traceable narrative" and succinct appeal to the "indexiability" of on-line auction sites:

> DIOR BRACELET: 100% authentic comes with original box. Box has a few dirty marks as I bought this 6 months ago. NEVER been used. Its beige in colour with gold metal. This pic isn't great and looks pinkish in colour but the item isn't. (eBay.co.uk/jewellery and watches/bracelets 16 July 2009)

To clarify the standing of the seller, it is not just seller ratings that are viewed (these are widely understood to be so normative as to be rendered meaningless) but the standing of the particular item. By tracing other items presently for sale by the seller and those previously sold by the seller, the potential bidder can piece together a constellation of brands and other objects; indeed an entire material culture profile that feeds-back as a testament to provenance that helps confer value on the listed article. The Dior bracelet mentioned previously, for example, is enmeshed in a series of other sales, across eBay's temporal spatiality, that offers insight into the seller's size, brand preferences, and even tastes in cosmetics. Clearly, this provenance also casts light on whether or not items actually ever belonged to a specific seller or whether they are stolen goods (as might be the case in terms of packaged branded cosmetics and easily "shop-liftable" items) (see Masnick 2008).

Wengrow (2008) has argued that in understanding branded commodities too little attention is paid to the actual processes of commodity marking, emphasis being placed instead on the transformative power of social relations to alter the status of objects. On eBay, the mark of the brand is reduced to its suitability to on-line visual rhetoric—that which can be captured of the brand in ostensibly amateur digital photos and documentary evidence. Boxes, original packaging, and original labels are vital, not just in authenticating the branded object but in authenticating the seller him or herself. In the sales narratives, attention is carefully drawn to any subjective physical interventions (dirty marks on the box, infants' dribbles on the car-seat straps) that compromise the objective marks of the branded commodity (be it a Dior bracelet or a Max-Cosi car seat).

It is the coming together of brand and narrative testament and the value generated through the marks of use that makes the archetypal on-line sale. The brand as used—or leased—carries an alternative: non-first-purchase value; freed from the mechanisms of retail and advertising, it is perhaps unsurprising that many eBay browsers use this "archive of everyday material culture" as a pooled guide to brand salience. Brands, as Arvidsson has observed, are based on "values, commitments and forms of community sustained by consumers." They are also "mechanisms that enable a direct valorization (in the form of share prices, for example) of people's ability to create trust, affect and shared meanings" (2005: 236). In this "interstitial site of material and immaterial cultures" (Hillis 2006), the idea of the brand is not confined to contemporary branded goods circulated and managed in the first-sale arena.

Ercol, a British postwar furniture brand renowned for its quality and conservative interpretation of modernist design, is a form of material culture that lends itself perfectly to the on-line auction format. The popularity of Ercol vintage furniture and the subsequent rise in its collectable value is tied to its branded style commodity and the ease with which its homogeneous forms are recognised and authenticated through eBay archival. That the moderately priced elm furniture has been a staple of middle-class British householders since the 1940s means that there is the volume of objects to sustain a brand identity, rather than generate just a collectible niche.

Richard, 34 (IT consultant, South London), initially used eBay when it was recommended as an efficient means of sourcing a spare part for his aging Peugeot. After two successful bids and purchases for car spare parts, Richard became, in his own words, "an Ercol watcher." Richard stumbled across the Ercol furniture when browsing eBay for a functional bookshelf—though generally uninterested in collectable or vintage furniture, he has subsequently purchased and sold a dozen Ercol items through eBay. He said:

> I've never been to a real auction as such; it's just not something I'd do—way too scary! But on eBay—well to be honest I've decked out the whole flat using eBay—I know that if I say, buy a piece of branded vintage furniture—with all the marks— if it doesn't fit—I just put it straight back on and get my money back or maybe even make a profit.

Once the staple of local household clearance and junk shops, Ercol has evolved as an archetypal on-line U.K. auction brand: Its items are homogenised into types; its crafted smooth elm surfaces and patina are suited to the visual rhetoric of on-line listings; and the items bear a maker's label. Their provenance is traceable and their models and prices can easily be compared, at the press of a button, with an index of previous buyers and previous prices. It is the virtual translation of the furniture's materiality and the legibility of its trademark that has transformed Ercol into what is effectively a premium second-hand brand.

Only the owner of the winning bid experiences the material object. eBaying, as a cultural practice, more typically involves the repetitive, serialised consumption of eBay objects as images and virtual encounters. This transitory mode of distribution matches what Gregson (2007) describes as the increasing normativity of transience as regards the practice in accommodating certain types of our things. Despite the promotion of eBay as a forum of memory and collectibles, as an archive of everyday material culture, it thrives on a recognised temporality of branded goods. eBay is widely recognised as having originated as an arena for selling out-moded technology. Although Ercol furniture sustains a static branded archival presence, for contemporary branded products it is also the site in which obsolescence and brand depreciation is effectuated.

Caroline bought her Bugaboo stroller (with accessories) first hand at the cost of £800 2 years ago, and, along with a Danish STOKKE cot, Baby Bjorn bouncy chair, and a Maxi Cosi car seat she has carefully stored all the instruction manuals, guarantees, and original packaging so that when she comes to resell them on eBay, she will re-assemble them as branded commodities. She also keeps the boxes from her branded children's toys (such as Playmobil, Lego, etc.), knowing that she will be releasing them back onto the market. Pre-emptive brand resale, underpinned by the traceability and indexibility of on-line auction, thrives in the fluid baby and children's wares market. To quote Caroline:

> Anything that is branded, basically, I see weirdly now as a bargain. I mean home-knitted kids' jumpers and expensive hand-made dolls houses are lovely but they're going nowhere on eBay. You need a brand that people know works well. It's like a great big pool of kids stuff and we're all just borrowing it and putting them back again.

eBay-branded goods take on an inalienable status (they are cherished, protected, and maintained—kept in the living collective brand archive of on-line auction), and the pre-emptive resale asserts a backward influence over the types of branded goods suited to the virtual second-hand market. A successful eBay trader of children's wear estimated that some of the branded articles she purchases for resale in her virtual shop had been through as many as 12 eBay transactions by the time she purchased them. As she said: "They may not even be that worn, because the mothers' have kept them like-new just so they can resell easily."

The brand managers of child-related products such as Bugaboo and Maxi-Cosi have responded to the threat of the second-hand liquid asset approach to their goods by generating increasingly diverse product lines that make their products more individuated and therefore less standardised for second-hand sale (e.g., items in a tiger pattern, paprika pink, or primrose yellow). Previously functional products such as children's car seats have joined the 'It' accessories of Bag, Steal and Borrow in their nuanced relation to changes in fashion and style and their appeal as "liquid assets."

Conclusion

In her discussion regarding the discrepancy between legal/global corporate definitions of authenticity pertaining to product copyright and value, and locally generated product cultures, Vann (2006) contrasts the official definition of authentic products as those "whose brand names and logos are said to signify truthfully and accurately a specific corporate origin" (p. 287) with those practised by Vietanamese consumers on an everyday basis. Direct brand copies, for example, are considered unproblematic in Vietnam as mimic goods that usefully match a new middle-class requirement for cheaper but "branded" goods.

Similarly, in the simultaneously local and global marketplace of eBay, formal "official" primary market–led definitions of branded commodities are not simply grafted into the definitions of "brandedness" generated in the various cultures of the second-hand market. The rise in the transient possession of brands is not, as some of the marketing literature suggests, simply the result of a rational economic consumer applying themselves to the newly rationalised territory of the post-primary market. In the processes of trading-up, branded commodities are authenticated through "pre-loved" haptic worlds and narratives that lend a provenance combining formal sanctioning with a form of savvy investment in the brand. The enduringness of the brand in the virtual space of eBay's perpetual archive allows for a transient version of possession that is tied to an unprecedented abundance of commodities and the demise of pre-on-line face-to-face divestment practices.

Although certain types of branded commodities, such as cars, have long been understood in relation to their depreciation and second-hand brand appeal, the boom in the secondary market is reconstituting the very categories that branded commodities inhabit. Definitions of luxury and designer, once tied to the materiality and skills of related industries from couture to leather goods, are now as relevant to functional baby goods as they are to objects of high-end fashion. As brand managers try to engineer the temporalities of their goods, generating obsolescence through the introduction of ever-faster design changes, it is in the consensual forum of the mass second-hand market that increasingly negotiates and dictates this process. Designers are still, to quote Papanek (1985 [1971]) the "handmaidens of capitalism," but it is the alternative economies of disposal that increasingly steer the pace of brand change and the aesthetic and style intonation.

Critics of second-hand on-line auction sites, such as Zukin (2005), take issue with the cyber-auction conjecture that everything in contemporary culture, from family heirlooms to pre-worn branded sneakers, is up for sale. But her contention, as Desjardins points out in her study of ephemera on eBay, denies the access to alternative knowledges and information that eBay allows (2006). As brand dispersal and realisation have become increasingly domesticated, brands, rather than being neutralised in the second-hand setting, have become key agents in circumscribing the valourisation of social processes ranging from the memorialising of kinship to the constituting genres of parenting.

At first glance, the proliferation of a "lease" and "everything up for sale" economy seems indicative of a Baumanesque (Bauman 2003) "liquid love" model of relationships in which the fleeting value of goods matches the superfluity of contemporary social relations. However, it seems that babies are not any less valued because their brand-marked cribs and strollers are circulated and revalorised in the collective space of eBay rather than kept in the family attic to gather dust.

Notes

1. According to the Nissanof study, eBay is estimated to have 180 million traders worldwide and a turnover of $130 million in daily transactions.
2. In the growing business and marketing literature addressing the spectre of the second-hand brand market, the Bugaboo stroller is frequently cited as the prime example of the importance of brand strategists in trying to "manage" the renegade culture of "preemptive secondary market purchase." It is recognition of a new consumer approach to first-time purchase (an approach that prioritises and preempts secondary-market brand value) that has heightened the Bugaboo's investment in "design obsolescence" (the use of novel fashion colour palettes and customisation) as a means of differentiating between primary and secondary markets. Madeleen Klassen, Bugaboo's international marketing director, describes the spectre of the "liquid asset," saying: "The so-called lease economy is getting clearer thanks to the eBays of this world. The Internet is very visible. You can easily follow products moving around." (Cuddeford Jones 2006 : 3)

   Bugaboo, a pram whose popularity has been tied to its media profile as the must-have item of celebrity mothers, is a self-consciously innovative design emanating from the conceptual Eindhoven Design Academy in the Netherlands. Its radical design positioned it outside the mainstream traditional baby-wares mass market, and, as such, it began as a relatively obscure product with niche market appeal. A high-end product item, Bugaboo's sales have grown in line with the market growth of eBay, which has provided the ideal forum to resell at a measurable, calculated value. Although the second-hand economy initially threatened to undermine the product's first-hand market, the brand managers of Bugaboo now incorporate the "lease" economy approach into their design and pricing concepts; to quote Klassen, "Consumers are reassured a resaleable product and then know it is worth investing in our brand." A fast turnover of new designs and colour schemes has become the only means of differentiating first phase models from their second-hand counterparts.

3. Tina Brown, leading international fashion editor, identified "fabloids" as a distinct genre of publication interweaving celebrity gossip, fashion, and luxury brands that "combines the tabloid hunger for sensation with the requirement to always look fabulous" (Brown 2007: 337, quoted in Chapter 11, "Contemporary Glamour," Gundle (2008: 347–87).
4. Unlike other product industries, the car industry and its brand managers have always incorporated the resale and depreciation value into their brand strategies and calculations of brand equity (see, e.g., Betts and Taran 2004 and Purohit 1992). However, it is the *cultural* after-life of the brand that this anecdote suggests is being so vigilantly controlled by brand managers.
5. *Futureshop* (Nissanoff 2006), selected by the *Wall Street Journal* as one of the five best books on consumer culture, belongs to a growing business literature and media acknowledging the impact of on-line auctions and second-hand markets. See, for example, Adler (2002), Bunnell (2000), Cohen (2002), and Helft (2001).
6. Walpole British Luxury protects the intellectual property rights of British luxury goods companies including Jimmy Choo and Links jewellers. LHMV is a luxury group representing 60 brands, including the fashion labels Louis Vuitton, Dior, and Fendi.

References

Aaker, D. 1991. *Managing Brand Equity.* New York: The Free Press.

Aaker, D. 1996. *Building Strong Brands.* New York: The Free Press.

Adler, J. 2002. The eBay Way of Life. *Newsweek*, 17 June, pp. 50–60.

Arvidsson, A. 2005. Brands: A critical perspective. *Journal of Consumer Culture* 5:235.

Baudrillard, J. 1996 [1968]. *The System of Objects.* London: Verso.

Bauman, Z. 2003. *Liquid Love: On the Frailty of Human Bonds.* Cambridge: Polity Press.

Betts, S. C. and Z. Taran. 2004. The "brand halo" effect of durable goods prices: Brand reliability and the used car market. *Academy of Marketing Studies Journal* 8(1): 7–18.

Brown, J., A. J. Broderick, and L. Nick. 2008. Word of mouth communication within online communities: Conceptualizing the online social network. *Journal of Interactive Marketing* 213:2–20.

Brown, T. 2007. *The Diana Chronicles*. London: Century.

Buchli, V. and G. Lucas. 2001. The archaeology of alienation: A late 20th century British council flat. In V. Buchli and G. Lucas (Eds.), *Archaeologies of the Contemporary Past*: 158–67. London: Routledge.

Bunnell, D. (with R. A. Luecke) 2000. *The eBay Phenomenon: Business Secrets behind the World's Hottest Internet Company*. New York: John Wiley & Sons.

Campbell, C. 1987. *The Romantic Ethic and the Spirit of Modern Consumerism*. New York: Basil Blackwell.

Clarke, A. 2000. Mother swapping: The trafficking of nearly new children's wear. In P. Jackson, M. Lowe, D. Miller, and F. Mort (Eds.), *Commercial Cultures: Economies, Practices Spaces*: 85–100. Oxford: Berg.

Clarke, A., Ed. 2010. *Design Anthropology: Object Culture in the 21st Century*. New York: Springer.

Clarke, A. and D. Miller. 2002. Fashion and anxiety. *Fashion Theory* 6:191–214.

Cohen, A. 2002. *The Perfect Store: Inside eBay*. Boston: Little Brown.

Crewe, L. and N. Gregson. 1998. Tales of the unexpected: Exploring car boot sales as marginal spaces of contemporary consumption. *Transactions of the Institute of British Geographers* 23:39–53.

Cuddeford Jones, M. 2006. Consumption: Sale or return, *Brand Strategy Magazine* (July 2006).

DeSilvey, C. 2006. Observered decay: Telling stories with mutable things. *Journal of Material Culture* 11:318–38.

Desjardins, M. 2006. Ephemeral culture/eBay culture: Film collectibles and fan investments. In K. Hillis, M. Petit, and N. Epley (Eds.), *Auctioning the Authentic: eBay, Narrative Effect, and the Superfluity of Memory*: 31–43. London: Routledge.

Drazin, A. and P. Garvey. 2009. An exploration by way of an introduction: Design and the having of designs in Ireland. *Anthropology in Action* 16. *Special Issue Title: Special Joint Issue with the Irish Journal of Anthropology and Journal for Applied Anthropology in Policy and Practice* 16:4–17.

Edwards, C. 2005. *Turning Houses into Homes: A History of the Retailing and Consumption of Domestic Furnishings*. Aldershot, UK: Ashgate.

Fitzmaurice, J. 2008. Splurge purchases and materialism. *Journal of Consumer Marketing* 25:332–38.

Frank, T. 1997. *The Conquest of Cool: Buisness Culture, Counterculture, and the Rise of Hip Consumerism*. Chicago: University of Chicago Press.

Glaister, D. 2006. A brush with the law. *The Guardian*, 2 August, p. 14.

Gorz, A. 1980. *Farewell to the Working Class*. London: Pluto.

Gorz, A. 2003. *L'Immatériel*. Paris: Galilée.

Gregson, N. 2007. *Living with Things: Ridding, Accommodation, Dwelling*. Wantage, UK: Sean Kingston Publishing.

Gregson, N. and L. Crewe. 2003. *Second Cultures*. Oxford: Berg.

Grove-White, A. 2001. No rules, only choices? Repositioning the self within the fashion system in relation to expertise and meaning: A case study of colour and image consultancy. *Journal of Material Culture* 6:193–211.

Gundle, S. 2008. *Glamour: A History*. Oxford: Oxford University Press.
Hawkins, G. and S. Muecke. 2003. *Culture and Waste: The Creation and Destruction of Value*. Lanham, MD: Rowman and Littlefield.
Helft, M. 2001. What makes eBay unstoppable? *Industry Standard*, 6–13 August, p. 35.
Herrmann, G. and S. Soiffer. 1984. For fun and profit: An analysis of the American garage sale. *Journal of Contemporary Ethnography* 12:397–421.
Hetherington, K. 2004. Secondhandedness, consumption, disposal and absent presence. *Environment and Planning D: Society and Space* 22:157–73.
Hillis, K. 2006. A space for the trace: Memorable eBay and narrative effect. *Space and Culture* 9:140–56.
Hinton, J. 2002. By sale, by gift aspects of the resale and bequest of goods in late-sixteenth-century Venice. *Journal of Design History* 15:245–62.
Kleinman, M. 2008. eBay to hold "counterfeit summit" with Jimmy Choo and other British icons. *The Daily Telegraph*, 25 July, Finance Section.
Lemire, B. 2004. Shifting currency: The culture and economy of the second-hand trade in England, c. 1600–1850. In A. Palmer and H. Clarks (Eds.), *Old Clothes, New Looks: Second-Hand Fashion*: 29–48. Oxford: Berg.
Lucas, G. 2002. Disposability and dispossession in the twentieth century. *Journal of Material Culture* 7:5–22.
Masnick, M. 2008. Retailers blame eBay for driving good people to shoplift. *TechDirt*, 23 September. www.techdirt.com (accessed 29 March, 2010).
Miller, D. 1995. *Worlds Apart: Modernity through the Prism of the Local*. London: Routledge.
Modleski, T. 1986. Femininity as mas(s)querade: A feminist approach to mass culture. In C. MacCabe (Ed.), *High Theory/Low Culture: Analysing Popular Television and Film*: 37–52. Manchester, UK: Manchester University Press.
Nissanoff, D. 2006. *Futureshop: How to Trade up to a Luxury Lifestyle*. New York: Penguin Press.
Norris, L. 2004. Shedding skins: The materiality of divestment in India. *Journal of Material Culture* 9:59–71.
O'Cass, A. 2004. Fashion clothing consumption: Antecedents and consequences of fashion clothing involvement. *European Journal of Marketing* 38:869–82.
Papanek, V. 1985 [1971]. *Design for the Real World: Human Ecology and Social Change*. London: Thames and Hudson.
Park, H. J., L. D. Burns, and N. J. Rabolt. 2007. Fashion innovativeness, materialism, and attitude toward purchasing foreign fashion goods online across national borders: The moderating effect of internet innovativeness. *Journal of Fashion Marketing and Management* 11:201–14.
Park, H. J. and B. L. Davis. 2005. Fashion orientation, credit card use, and compulsive buying. *Journal of Consumer Marketing* 22:135–41.
Peiss, K. 1998. *Hope in a Jar: The Making of America's Beauty Culture*. New York: Henry Holt & Company.
Purohit, D. 1992. Exploring the relationship between the markets for new and used durable goods: The case of automobiles. *Marketing Science* 11:154–68.

Rathje, W. L. and C. Murphy. 2001. *The Archaeology of Garbage*. Tucson: University of Arizona Press.

Reno, J. 2009. Your trash is someone's treasure. The politics of value at a Michigan landfill. *Journal of Material Culture* 14.1:29–46.

Robins, K. and F. Webster. 1999. *Times of Technoculture: From the Information Society to the Virtual Life*. London: Routledge.

Rosencrance, L. 2008. eBay, British luxury goods group to discuss counterfeit sales: Meeting comes on heels of differing rulings on who bears responsibility for preventing sales of bogus goods. *Computerworld: Networking and Internet*, 24 July.

Scanlan, J. 2005. *On Garbage*. Chicago: University of Chicago Press.

Schmitt, B. H. 1999. *Experiential Marketing: How to Get Customers to Sense, Feel, Think, Act and Relate to your Company and Brands*. New York: The Free Press.

Smith, C. W. 1989. *Auctions: The Social Construction of Value.* London: Harvester Wheatsheaf.

Soteriou, H. 2007. Dismay at eBay's unsatisfactory response to "bid shielding" offences: Growth in high bids which vanish sparks concern among auction site users. *The Guardian*, 17 May.

Stallybrass, P. 1998. Marx's coat. In P. Spyer (Ed.), *Border Fetishisms: Material Objects in Unstable Spaces*: 183–207. London: Routledge.

Stalnaker, S. 2002 Hub Culture: The Next Wave of Urban Consumers, Singapore: John Wiley & Sons.

Stewart, M. 1992. Gypsies at the horse-fair: A non-market model of trade. In R. Dilley (Ed.), *Contesting Markets: An Analysis of Ideology, Discourse and Practice*: 79–96. Edinburgh: Edinburgh University Press.

Vann, E. 2006. The limits of authenticity in Vietnamese consumer markets. *American Anthropologist* 108:286–96.

Wengrow, D. 2008. Prehistories of commodity branding. *Current Anthropology* 49:1.

Zukin, S. 2005. *Point of Purchase: How Shopping Changed American Culture*. New York: Rouledge.

About the Editors and Contributors

Editors

Andrew Bevan
UCL Institute of Archaeology
31–34 Gordon Square
London WC1H 0PY
United Kingdom

Andrew Bevan is a Lecturer in Archaeology at University College London. He has conducted fieldwork and related research in a wide range of locations, but with a core regional focus in the eastern Mediterranean. His primary interests include theoretical approaches to object value, early complex societies, landscape ecology, spatial analysis, and comparative archaeological and anthropological methods. Recent publications include a book entitled *Stone Vessels and Values in the Bronze Age Mediterranean* (Cambridge University Press, 2007).

David Wengrow
UCL Institute of Archaeology
31–34 Gordon Square
London WC1H 0PY
United Kingdom

David Wengrow is a Reader in Comparative Archaeology at University College London. His research addresses the early development of complex societies in the Middle East and their implications for world history. He has a long-standing interest in anthropological approaches to art and material culture and their applications in archaeology. He is the author of "Prehistories of Commodity Branding" (*Current Anthropology*, 2008), which extends the study of branding practices back to the beginnings of urban life in ancient Mesopotamia. He is also the author of two books: *What Makes Civilization? The Ancient Near East and the Future of the West* (Oxford University Press, 2010) and *The Archaeology of Early Egypt: Social Transformations in North-East Africa, 10,000–2650 BC* (Cambridge University Press, 2006).

Contributors

Gracia Clark
Department of Anthropology
Indiana University
Bloomington, IN 47405-7100
USA

Gracia Clark (PhD, Cambridge) has studied the Kumasi Central Market, an urban daily market in Ghana with 20,000 traders, since 1978. She consulted for the International Labour Organization and the United Nations Development Fund for Women for several years before teaching anthropology at the University of Wisconsin, Parkside, the University of Michigan, Ann Arbor, and now Indiana University, Bloomington. She has edited *Traders vs. the State* (1988) and *Gender at Work in Economic Life* (AltaMira Press, 2003), written on relations of credit, leadership, and domestic work, and addressed development issues of food security, gender, and trade liberalisation. *African Market Women: Seven Life Stories from Kumasi* (Indiana University Press, 2010) complements her 1994 book *Onions Are My Husband* (University of Chicago Press).

Alison J. Clarke
University of Applied Arts
Design History and Theory
Oskar Kokoschka-Platz 2
A-1010 Vienna
Austria

Alison J. Clarke is a Professor of Design History and Theory and Research Director of the Victor J. Papanek Foundation, at the University of Applied Arts Vienna, Austria. She has previously written and researched widely on the design and anthropology of brands; publications include *Tupperware: The Promise of Plastic in 1950s America* (Smithsonian Press, 2000) and, most recently, *Design Anthropology* (Springer, 2010).

Magdalena Crăciun
Str. Rahovei Bloc A3 sc 2 ap 28
Turnu Magurele
145200
Romania

Magdalena Crăciun earned her PhD from the Department of Anthropology, University College London, in January 2010. Her doctoral research was a multi-sited ethnography of people's engagement with fake branded clothing on a periphery of Europe, in three distinct locations: Istanbul, a major producer of fake brands; Europa market in Bucharest, the main source of counterfeits in Romania; and a provincial Romanian town, as an ordinary place of consumption. She is currently preparing a book manuscript and several other articles on these topics.

Ferenc Hammer
ELTE Media Department
Múzeum krt. 6–8, 1088
Budapest
Hungary

Ferenc Hammer (PhD, ELTE University) is an Assistant Professor at the Institute for Art Theory and Media Studies at ELTE University, Budapest. He teaches and conducts research on artefacts, practices, and memories associated with everyday life. His recent publications include: "Sartorial Manoeuvres in the Dusk: Blue Jeans in Socialist Hungary," in F. Trentmann and K. Soper (Eds.), *Citizenship and Consumption* (Palgrave Macmillan, 2008); "Magyar Denim: Metamorphoses in the Clothing Factory," forthcoming in *Denim: Unravelling Contradictions* (a special edition of *Textile: The Journal of Cloth and Culture*, June 2010).

Marcos Martinón-Torres
UCL Institute of Archaeology
31–34 Gordon Square
London WC1H 0PY
United Kingdom

Marcos Martinón-Torres is a Senior Lecturer in Archaeological Science and Material Culture at the University College London. With degrees in history, archaeology, and archaeological science, he has a special interest in the integration of different research approaches to past materials and technologies. Much of his research has focused on medieval and post-medieval metallurgy, with a strong emphasis on the study of al/chemical and metallurgical laboratories and the processing of noble metals. He is also interested in the transfer of material culture and ideas between indigenous peoples and Europeans in colonial contexts. He currently works on projects in Europe, America, and China.

Rosana Pinheiro-Machado
Protásio Alves, 1296/702
90410-005
Porto Alegre/RS-
Brazil

Rosana Pinheiro-Machado is a Lecturer in Anthropology at the Superior School of Advertising and Marketing (ESPM), Porto Alegre, Brazil, with a PhD in social anthropology from the Federal University of Rio Grande do Sul, Brazil. She specialises in the ethnography of Chinese and Latin American societies and has been carrying out multi-site fieldwork in both of these regions over the last 10 years. Previous topics on which she has published include labour practices, social inequalities, capitalism in emergent economies, *guanxi*, informal economies, kinship and migration, legality and illegality, human rights, globalisation, fakes, and brands.

Jean-Pierre Warnier
32, rue de l'ERMITAGE
91700 Ste Geneviève-des-Bois
France

Jean-Pierre Warnier (PhD, Anthropology, University of Pennsylvania, 1975) taught anthropology in Nigeria, Cameroon, and, since 1985, at the University Paris-Descartes. He is now a Professor Emeritus and associated researcher at the Centre d'Etudes Africaines, Paris. His books on political economy in Cameroon and on bodily and material culture include *Echanges, Développement et Hiérarchies dans le Bamenda Pré-colonial—Cameroun* (Franz Steiner Verlag Wiesbaden, 1985); *The Pot-King. The Body and Technologies of Power* (Brill, 2007); and *Régner au Cameroun. Le Roi-Pot* (CERI-Karthala, Coll. "Recherches internationales," 2009).

Jason Yaeger
Department of Anthropology
One UTSA Circle
University of Texas at San Antonio
San Antonio, TX 78249
USA

Jason Yaeger is an Associate Professor of Anthropology at the University of Texas at San Antonio. Since 1990, he has conducted archaeological fieldwork in Belize, Bolivia, the United States, and India to address his interests in the archaeology and ethnohistory of Mesoamerica and the Andes, the organisation of ancient households and communities, ancient urbanism, and material culture and identity. He has published *Classic Maya Provincial Politics: Xunantunich and Its Hinterlands* (with Lisa J. LeCount, University of Arizona Press, 2010) and *The Archaeology of Communities: A New World Perspective* (with Marcello A. Canuto, Routledge, 2000).

Index

D

O

P

R

S

T

U

V

W